Raising Ducks

A Beginner's Guide to
Raising Healthy and Happy Backyard Ducks

By: Irene Mills

Table of Contents

Chapter 1: My Duck Story - Dreams vs. Reality

Duckling fever had gotten the best of me.

I was on my way home from Tractor Supply Company, barely able to contain my excitement about the fuzzy babies that snuggled on my lap. Warnings I had received about messy ducks flew out the window as I sped along the highway. There was no mess that could diminish the delight of that moment. This was going to be worth every inconvenience, guaranteed.

I had raised ducks before when I was a child. Sure, we had gotten rid of them when they were teenagers because they were so messy. But I was confident I could handle the work of caring for them now that I was an adult. I had raised chickens for almost 28 years. Ducks were no match for me.

As I cuddled the fuzz balls, I let my fairy-tale wish dreams run rampant. I wished for them to become snuggly, cuddly, and close. I wished for them to be every bit as tame and lovable as the chickens I had raised since babyhood. I wished for them to follow me around the yard. I wished for them to drink clear, cool water under a mulberry tree like the ducks in one of my favorite books, *Angus and the Ducks*. They would delight my duck-loving nephew as they waddled around the yard. My dream was complete, and I had no reason to doubt.

During the ducks' first few hours, they were every bit as snuggly and entrancing as I had hoped. They peeped softly, nuzzled their way into my jacket, and crawled adorably up my jacket to snuggle under my hair on my shoulder.

As the weeks went by, I was powerfully reminded of just how messy ducks can be. Every few hours, the ducklings managed to soak the newspaper in their brooder box with water and manure. Three or four times a day, I was gathering up soggy brooder lining and carrying it to the trash can. My office room where the ducklings resided now carried a distinctive odor. Replacing their bedding and running an air purifier wasn't helping much. How could two very small creatures be so messy?

In no time, I began to realize that the ducks would provide a great parable for the video series I was creating. The pure white Pekin ducks had moved outdoors, and I had planted a wonderful garden of mulberry trees, wheat, alfalfa, and clover for them. The plants were establishing themselves underneath the raised mesh screens I had built. In this way, the ducks could walk on top of them, splash water down on the seedlings, and drop fertilizer through the mesh onto the seedlings. The only problem was the ducks provided so much manure that they suffocated and exterminated all the plants.

My fairy-tale wish dreams were going out the window. Drinking water under the mulberry trees looked more like splashing water all over the cage and then wallowing in manure and mud. Snuggling and cuddling and following me around the yard looked more like running from me as if I were a terrorist and being way too messy to even think of holding.

Just take a look at these photos. And remember, they are pure white when they're clean.

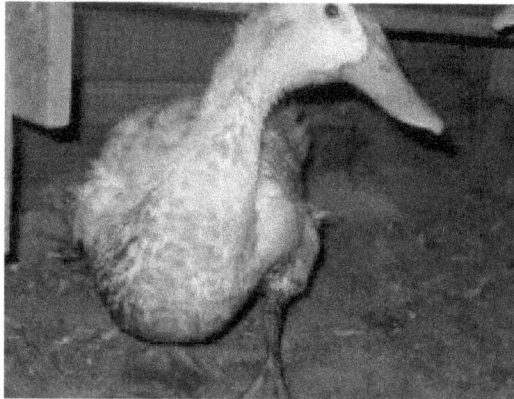

Ducks enjoying the perks of mud.

Just like the ducks, relationships don't always turn out the way we expect. Situations are messier than we thought they would be. People don't read the script of our expectations and live them out as if we were the director and they were the actors. Even our faith doesn't always agree to follow *our* plan.

Ideas took shape in my mind. I would create a lovely video including shots of my ducks playing in the mud, quacking as I gave them food scraps, and splashing in their pool. I would discuss how we can learn to enjoy relationships, even when they don't meet our previous ideals. Instead of detaching from people when they don't live up to our fairy-tale script (just like I don't ditch my ducks for their mess), and instead of forcing them to be what we'd hoped they'd be (just like I don't chase my ducks around requiring them to let me hold them), we can simply enjoy them for who they are. Live for the moment. Enjoy what *is*.

As I finalized my plans for the video, a young friend of mine agreed to be the narrator for the video. She was over-the-top excited to be included and thought that the concepts were interesting and inspiring. We started meeting together as we began the process of creating the video.

I had all my ducks in a row.

Finally clean... for a few minutes!

But then the plot thickened.

As winter progressed, I spent a lot of time trying to create a watering system to meet the ducks' unique needs for plenty of fresh, ice-free water. I shopped all over town for equipment and later returned half of it. I exerted a lot of time and muscle on a 55-gallon watering system which I later had to abandon. I purchased a new watering system and tried to determine which stock tank deicer would operate with our electrical system without risk of electrocution for myself or the animals.

Despite my best efforts, I had to battle chunks of ice. I felt like a work horse as I hauled fresh warm water from the house, gallon after gallon, load after load, day after day. The specialized hose I had purchased broke while I was out of town and the ducks were with a caregiver.

Worst of all, the ducks had grown to an enormous size and were creating more manure than I had ever dreamed possible. I had cleaned chicken coops most of my life, but duck coops were a different story. Sacrificing every Saturday morning to scooping unmentionable sizes and consistencies of manure (and living, wiggling critters growing in the manure) was beginning to make me doubt my sanity.

One bitterly cold day, I could not ignore the mess one minute longer, despite the fact that I was sick with a cough and fever. As the temperatures dropped, so did my determination. Shoveling robotically with a weary mind, every dream for my ducks, the videos, and the truth they represented slowly drained from my glazed, feverish eyes.

To complicate matters, the ducks hated being photographed. Every attempt to accustom them to a video camera failed. They would run from me if I came near them with even the smallest of cameras. How would they let a professional filmmaker with lots of equipment even get near them?

(Here is the view I got when I was trying to take some photos to demonstrate my plans for the project. No, she was not laying an egg, and no, I was not torturing her. She just didn't want to be coerced into cooperating with my schemes.)

No one's going to tell this duck what to do!

Most importantly, the girl who had agreed to be my narrator quit, making it difficult for me to carry on with plans for the project. I realized I had encountered a much bigger lesson on expectations than I had ever dreamed. For now, I would have to realize this idea was out of my own hands, and channel my creativity in other directions.

As soon as the ducks saw that I was off their case and the camera was put away, they became much more amiable. They started following me around like I had always wished. Every morning, they met me with their loud quacking and entertained me frequently with their crazy antics.

Just because we don't have the relationship I hoped for doesn't mean we can't have a relationship. It may not include cuddling and closeness, but our relationship does include a lot of fun and enjoyment. I love watching them bob their heads and dance in excitement about (*wait for it...*) the fact that they are drinking water!! Just watching them enjoy their life is a reward.

Ducks are strange and crazy and eccentric and full of personality and suspicious and loud and opinionated and playful and entertaining and personable. They are the most laughter-inducing pet I've ever had. (Just ask any duck owner on the online duck-owner support group! I bet the laughter emoji and the laugh with tears emoji and roll on the floor laughing emoji are used more in that group than any other backyard animal group.)

Ducks and relationships are messy, but they are always worth the effort. Beyond a doubt, ducks are the most hilariously worth-it creatures I've ever owned. And beyond a doubt, every human being is a delightful investment worth every inconvenience we may suffer.

As the years went by, I learned some tips and tricks to make living with ducks enjoyable, rewarding, and (dare I say it?) easier. And yes, I did finish up the video series. My ducks never did make it into the footage, but I co-opted some zoo ducks and some brand-new ducklings to fill the roles.

Over the years, I made many mistakes, cried some, laughed a lot, and learned some important lessons! Raising ducks is not for the faint of heart, but it is certainly a rewarding journey.

This book exists to help YOU decide if ducks are right for you. If you decide to purchase ducks, this book will help you avoid the mistakes that I made. It will give you the tools you need to raise healthy and happy backyard ducks. Let's dive in!

Chapter 2: Are Ducks for You?

Ducks are not for everybody. They are certainly not for the faint of heart. First, we'll look at a few reasons you may want to choose another farm animal. Then, we'll discover reasons you may want to add ducks to your backyard farm.

Let's get the bad news out of the way first.

If you want an easy, fun pet that requires little maintenance... then ducks are not for you. Are you looking for a pet you can stick in the backyard, occasionally refilling their food and water bowls while enjoying the feeling of being a hip homesteader? Are you hoping for the joy of looking out the back door and seeing a happy flock of poultry walking around and filling your heart with satisfaction?

Maybe you're hoping for a fun and easy pet for your kids. How about a bathtub buddy for your little ones? Here's what one brand-new duck owner had to say on her first week of duck owning: "I see lots of fun bath times in the future."[1] She posted the quip alongside photos of newly hatched ducklings being snuggled with her child. Her boy, discretely dressed in a bathing suit, was staring serenely at the newest member of their family as the two of them rested atop a crystal-clear bathtub. It took all my might to not to post a "laughing" emoji on the post. I wanted to write a couple days later, "Are the bath times still fun?"

If "easy" or "fun" are in your mind, stop reading this book right now and buy the *How to Raise Happy and Healthy Chickens* book. Maybe rabbits or goats. Anything but ducks.

If you hate manure, poop, and hard work... then ducks are not for you. There's no way to say this nicely. Ducks are MESSY. They're messier than you ever dreamed. Don't give the excuse that you know what it's like to clean up after a dog, cat, chicken, rabbit, or goat. Duck poop is thin, watery, and sticky. It gets EVERYWHERE! Ducks do not scratch their bedding like chickens do, so their poop accumulates on the top of the bedding. One duck owner described it like this: "Duck poop isn't like chicken poop. It's very wet. It's... *bloppy*. It just blops all over the place. Blop, blop, squirt, blop. That's ducks."

---

[1] Tammy King, Facebook post, March 13, 2021.

She goes on, "Duck poop sits on top of most coop litter. Then, big wide duck feet come along and trample it into a uniform coating of shit. It's as terrible as it sounds. If you keep ducks, all by themselves, on wood shavings or straw, you get something like this: a bottom, thick layer of perfect, dry straw, a middle layer of damp straw, and an upper layer of poop. Like poop icing on straw cake."[2]

Poop icing. That's exactly right. As I'll discuss later in this book, I started out raising my ducks and chickens together. Later on, I separated my ducks from my rooster because he got unexpectedly aggressive toward them. I quickly realized just how horrifying it is to keep ducks alone without proper bedding. That's why it's important to have the right kind of bedding if you're going to keep ducks. But more on that later...

If you want to keep an organized, clean, dry coop... ducks are not for you.
Two words that are synonymous with ducks are water and mud... oh yeah, and MESS. You've always known that ducks like water. But you probably have not yet fully understood the great extent of their absolute and total fascination, obsession, infatuation, and addiction with water. Ducks have a delusional craze for water. Just watch them dancing, stamping, and going crazy when you give them a new, crystal-clear pool full of water, and you'll understand how much they love it. Then, just stand by and watch as they shovel mud into it, poop in it, splash half of it out of the container within five minutes, shake it over the entire, nicely cleaned yard, and otherwise defile the entire pool you worked so hard on. Most of the time, you will need to change your ducks' water every single day. They will soil it within minutes, sometimes seconds.

And don't get me started on mud. My earlier story illustrated just how difficult the water and mud aspect can be for a new duck owner. As the ducks spill, slosh, spray, and track water everywhere, the floor of their cage can become mud. Yes, there are ways to manage the mud. Yes, there are tools you can learn to eliminate some of the disastrous slime. However, you'll have to do your research and put in the time and effort to eliminate mud. If you are obsessed with cleanliness, ducks are probably not the best option for you. Try a parakeet.

If you have an insecure personality who needs to be constantly affirmed... ducks may not be for you.
When raising ducks, you have to be ready for anything. Your ducks might become the cuddliest, sweetest pets on the face of the earth... or they may run from you as if you are a monster. Your ducks may swim in the beautiful, high-quality, amazing pool that you just spent thousands of dollars and hours of planning to create for them. Or they may refuse to even consider getting into it. Your ducks may love the types of foods that the Internet said they'd go crazy over. Or they may

---

[2] Erica, *Northwest Edible Life,* Last modified on May 17, 2017, Accessed March 29, 2021, https://nwedible.com/reasons-to-not-get-ducks/.

think you are giving them poison. I was in stitches over some of the stories that duck owners would tell over on our Backyard Ducks Facebook Group. Duck owners tell stories of buying the best and brightest for their ducks, only to have their ducks assume they are assassins. In other words, to survive duck ownership, you can't base your identity on your ducks' gratefulness. If your need for affirmation and reward is high, try something like a puppy.

If you just don't like to work... ducks are not for you. Ducks include hard work. Period. You will be shoveling manure, hauling water, cleaning out mud and muck, and satisfying ducks' infinite needs for food, water, attention, and more. There's no way to get out of the hard work, so you might as well get used to it now.

If you're hoping for a type of animal that only requires some food and water refills here and there and an occasional cage change, ducks are not the way to go. You might want to try a hamster.

If you love peace and quiet... ducks are not for you. Do you like sleeping in on a Saturday morning? Do you enjoy sitting in your backyard gazebo, soaking in the perfect silence and tranquility of a warm summer day? Then a pair of ducks may burst your bubble. Ducks will awaken you before dawn with their loud, insistent demands. These quacks are not the peaceful, loving contented noises you may imagine from ducks on a pond. They're not the gentle quacks you might have listened to on an idealistic "Sounds of Nature: Peaceful Ducks on a Lake" relaxing white noise video. Instead, their cries sound more like a yippy dog crossed with a screaming toddler. They are loud, insistent, and demanding. They will quack every time you go out of the house, every time you walk to your car, every time you talk in the kitchen, or every time you sing in the shower. They love quacking at anything unusual that is going on in their environment. They're not afraid to quack to say hello, to sound an alarm, to voice their approval, or to express their disapproval. If you post an innocent question on a duck Facebook group—something along the lines of, "How to help my ducks be quiet in the morning—I want to sleep!!" you are likely to get a lot of laughter and not much sympathy. The ideas include getting up and feeding them, stewing them for dinner, and putting in ear plugs. In short, almost all breeds of ducks are very noisy. I've heard that Muscovy ducks are also quite a bit less noisy than other breeds, although I've never raised Muscovy ducks myself. However, if you're looking for a quiet pet, you might want to check out another one of our e-books, *How to Raise Happy and Healthy Rabbits*. Or you could try a fish.

If you have issues with control... ducks are not for you. Ducks will make you feel like you are not in control. They are! They are demanding, entitled, and will let you know about it at every turn. They will inform you that you are their servant, and that they are the new kings. If you don't fulfill their wishes, they will let you know about it... loudly! If you tend to get irked by animals (or people) who demand things from you, don't give you space, or refuse to leave you alone, you might want to try a cat.

If you've read this far and are still interested in getting ducks, good for you! Ducks are funny, hilarious, entertaining, and rewarding. There are many excellent reasons to add ducks to your menagerie. Let's look at a few reasons you might want to buy some ducks this spring.

If you want a reason to get outdoors every day... buying ducks will be an excellent option for you! Are you looking for a motivation to get off the seat of your pants and get out into the great outdoors? Every single day? Without an excuse? A duck is a great motivation to do just that. If your memory needs jogging, your duck will remind you of your commitment. Better than any reminder on an exercise app, your ducks will break into your sleepy consciousness, drag you out of bed, and get you outdoors where you can truly thrive.

If you and your family need a productive outlet for exercise... buying ducks will be an excellent option for you! Ducks provide an excellent, meaningful way to exercise. As you'll learn in this book, shoveling the ducks' cage doesn't have to be a stinky, slimy chore. When you've managed your duck's bedding in a helpful and profitable way, cleaning the coop is a useful way to gain strength in your upper body. As you move bedding around, you exercise your arms and core. Your legs and core are challenged as you carry straw bales, move and pour bags of feed, and chase ducks into the coop at night.

If you need sensory stimulation for yourself or a special-needs child... buying ducks will be an excellent option for you! If you have a child with special needs, ducks may provide just the outlet they need to integrate their senses. Children with autism or sensory processing disorder need to do what therapists call "heavy work."[3] Any type of pushing, pulling, or carrying heavy objects will help meet their proprioceptive sensory needs. Ducks provide plenty of opportunity for deep pressure, carrying heavy items, and working against friction. It literally feels good—even for us as adults. When we've been sitting all day long at our office desks, spending time outdoors with ducks can be an excellent way to destress physically. All of us can benefit from the sensory stimulation of caring for ducks.

---

[3] Amanda Morin, "Heavy Work and Sensory Processing Issues: What You Need to Know," *Understood,* Accessed March 29, 2021,
https://www.understood.org/en/learning-thinking-differences/child-learning-disabilities/sensory-processing-issues/heavy-work-activities.

If you need a wonderful source of compost for your garden... buying ducks may be an excellent option for you! Ducks provide perfect conditions for making compost. Duck manure, mixed with straw, pine chips, leaves, food scraps, and generous amounts of water, are perfect ingredients for rich, luscious compost. Ducks will generate liberal amounts of these substances every year, so you will have a never-ending supply for your garden. You may even be able to sell compost on the side. In addition, duck manure can be applied instantly to your garden, whereas chicken manure must be seasoned for around a year before it can be used.[4]

If you need a source of natural anti-depressants in the compost... buying ducks may be an excellent option for you! You may notice that despite all the inconveniences of raising ducks, you get a "high" every time you're around them. Did you know there is a real, scientific reason for that? As we mentioned above, ducks create plenty of compost. And recent scientific discoveries have demonstrated that duck compost, just like any other kind of compost or rich soil, releases microbes that are good for your mental health. According to the *Atlantic*, compost provides a microbe that serves as an antidepressant.[5] The article explains, "M. vaccae, a living creature that resides in your backyard compost pile, acts like a mind-altering drug once it enters the human body, functioning like antidepressant pills to boost your mood."[6] You'll start noticing it right away: ducks make you feel good. Sometimes, when I'm tired and down, I just go outside and stand near the coop, watching the ducks' antics. In no time, I'm feeling better. The actual smell of the rich earth is reviving my spirit and mind with these helpful microbes. No wonder you can never bring yourself to get rid of the ducks, no matter how much they get on your nerves!

If you need a natural heater for your chicken coop... buying ducks may be an excellent option for you! Did you know that ducks' body temperature is higher than chickens? According to the Department of Animal and Food Sciences from the UK College of Agriculture, Food and Environment, chickens' body temperatures range from 105 to 107 degrees.[7] Ducks' body temperatures, however,

---

[4] Susan, "DUCK EGGS VS. CHICKEN EGGS: HOW DO THEY COMPARE?" *Tyrant Farms,* Last updated on January 13, 2019, Accessed March 29, 2021, https://www.tyrantfarms.com/5-things-you-didnt-know-about-duck-eggs/.

[5] Pagan Kennedy, "How to Get High on Soil," *The Atlantic,* Last modified on January 31, 2012, Accessed March 29, 2021, https://www.theatlantic.com/health/archive/2012/01/how-to-get-high-on-soil/251935/.

[6] "How to Get High on Soil"

[7] "Commercial Poultry Production Air Temperature," *University of Kentucky College of Agriculture, Food and Environment,* Accessed March 29, 2021, https://afs.ca.uky.edu/poultry/chapter-7-air-temperature.

average a degree higher, reaching up to 108 degrees Fahrenheit.[8] Because of this natural difference, ducks can help heat your coop in winter. Before I had ducks, my roosters would often get frostbite on their sensitive combs (wobbly red crests on their heads) and wattles (the lobes of flesh under their chins). Sometimes their toes would even fall off from the cold! Since I've gotten ducks, however, I've never had any problems with frostbite in my chicken flock. Rather than heating your chicken coop with dangerous, flammable heating options (and running up your heating bill), why not add some ducks to your flock? They are a safer, friendlier, and funnier heating option for your chicken coop.

If you want a source of delicious, hypoallergenic, oversized eggs… buying ducks may be an excellent option for you! One of the biggest benefits of raising ducks is their eggs. Ducks are prolific egg layers, and their eggs are large and delicious.

Duck eggs, in my humble opinion, are much more delicious than chicken eggs. The difference between duck eggs and chicken eggs mirrors the difference between sharp cheddar cheese and mild cheddar. Chicken eggs often have sharpness or a bite to them, but duck eggs are mild, rich, and delicious. For a person who gags on sharp cheese, duck eggs provide an excellent alternative.

In addition, duck eggs are a wonderful option for people with egg allergies. Susan from Tyrant Farms reports, "Since the protein in duck eggs is different than the protein in chicken eggs, many people with chicken egg allergies report that they're able to eat duck eggs with no problem."[9] After she discovered duck eggs, a friend of mine with multiple allergies was excited to once again enjoy omelets, scrambled eggs, and baked egg products. Duck eggs are very pricey, up to six times more expensive than chicken eggs.[10] So if you have allergies, buying yourself a couple of ducks can be a very positive option. They will provide you with a constant supply of delicious, hypoallergenic eggs.

Furthermore, duck eggs are larger than chicken eggs. These giant eggs go farther in omelets, scrambled eggs, and baking dishes. Ducks are prolific egg producers, producing a higher number of pounds per year, laying eggs for up to a year longer than a chicken, and producing eggs several weeks earlier than chickens.[11]

---

[8] "Winter Duck Care," *The Cape Coop,* Accessed March 29, 2021, https://thecapecoop.com/winter-duck-care/.

[9] Susan, "DUCK EGGS VS. CHICKEN EGGS: HOW DO THEY COMPARE?" https://www.tyrantfarms.com/5-things-you-didnt-know-about-duck-eggs/.

[10] Dan Nosowitz, "Everything You Need to Know About Duck Eggs," *The Modern Farmer,* Last modified on June 19, 2015, Accessed March 30, 2021, https://modernfarmer.com/2015/06/everything-you-need-to-know-about-duck-eggs/.

[11] Tyrant Farms

According to Healthline, duck eggs are even more nutrition-packed than chicken eggs.[12] Duck eggs contain high levels of iron, folate, and B12.[13] The high levels of folate make duck eggs an excellent choice for expectant mothers. Duck eggs contain high levels of healthy fatty acids, so they are a wonderful choice for health-conscious individuals.[14]

All in all, egg production is an excellent reason to get yourself some quackers!

If you have a giant backyard with a lake... buying ducks may be an excellent option for you. A big backyard is not a necessity for duck raising, by any stretch of the imagination. But if you're looking for a low stress option for buying ducks, you're in good shape. Having a large backyard helps with clean-up, since it gives the ducks a place to roam and more evenly distributes their excrement. A natural water source provides a huge benefit for you. It relieves you from having to change their water daily, haul water constantly, and work hard to provide clean, fresh bathing and drinking water for the ducks. A large yard with a pond or lake means that you can keep ducks with many less worries and much less work. Make sure your backyard is fenced.

If you want to teach your children the value of hard work... buying ducks may be an excellent option for you. In today's world, children are missing out on the important experience of working outdoors. Many kids spend their days relaxing in front of a screen—or multiple screens. As we mentioned above, too much time on screens can contribute to children becoming sensory deprived and or at higher risk of attention disorders such as attention deficit hyperactive disorder (ADHD).[15] Caring for ducks provides a chance for kids to get away from screens. It also provides important lessons in the value of hard work. It helps kids learn to stay focused and helps prepare them for adulthood.

If you love poultry but want a beneficial alternative to keeping chickens... buying ducks may be an excellent option for you. In many ways, ducks surpass chickens as ideal egg layers.[16]

---

[12] Cecilia Snyder, MS, RD, "Duck Eggs vs. Chicken Eggs: Nutrition, Benefits, and More," Last modified on December 1, 2020, Accessed March 30, 2021, https://www.healthline.com/nutrition/duck-eggs-vs-chicken-eggs#nutritional-comparison.

[13] Snyder

[14] Nosowitz, https://modernfarmer.com/2015/06/everything-you-need-to-know-about-duck-eggs/.

[15] "Preschooler Screen Time Linked to Attention Problems," *Cleveland Clinic,* Last modified on July 18, 2019, Accessed on March 30, 2021, https://newsroom.clevelandclinic.org/2019/07/18/preschooler-screen-time-linked-to-attention-problems/.

[16] https://www.tyrantfarms.com/a-fowl-battle-ducks-vs-chickens/

First, ducks do not scratch up your yard and garden.[17] If you've kept chickens for any length of time, you know that they have the pesky habit of demolishing gardens, newly planted seedlings, and basically any tender young plant in your yard—even and especially the ones you worked hardest to get established. Chickens displace your mulch, scattering it irretrievably around the yard. Chickens scratch dust and sand into the water bowl, which means you have to replace it more frequently. On the other hand, ducks walk around placidly with their soft, webbed feet. No more scratching!

Second, as we've already discussed, ducks lay larger eggs, begin laying earlier, and lay longer into old age.[18] Ducks can produce up to 32 to 52 pounds of eggs per year, whereas chickens only produce 22 to 34 pounds of eggs per year.[19]

Ducks are generally healthier than chickens and are more resistant to disease.[20] Ducks tolerate the cold better than chickens; and of course, they tolerate rain![21] You don't need to worry about your ducks standing outside on a cold, rainy day, shivering and literally freezing to death. According to Tyrant Farms, ducks only have a zero to three percent mortality rate per year, whereas chickens have a five to twenty-five percent mortality rate per year.[22] Ducks are able to gain a lot more nutrition from the natural environment than chickens can.[23] By slurping up worms, bugs, grass, and digging underground, they can feed themselves and contribute to their own nutrition.

If you need a natural way to eliminate bugs, pests, and weeds from your backyard... buying a duck may be an excellent option for you. Do you have a difficult time keeping your backyard free of weeds and overgrowth? Ducks love to graze and are good at keeping grass trimmed.

Perhaps your backyard is overrun by mosquitoes, flies, and other pests. Ducks would love to help you keep the pests under control. Sifting through the water with their excellent duck filters, ducks will help you take care of mosquitos in larval form.[24] Ducks also eat

---

[17] https://www.tyrantfarms.com/a-fowl-battle-ducks-vs-chickens/

[18] https://www.tyrantfarms.com/a-fowl-battle-ducks-vs-chickens/

[19] https://www.tyrantfarms.com/a-fowl-battle-ducks-vs-chickens/

[20] Jianmei Yang, Hongrui Cui, Qiaoyang Teng, Wenjun Ma, Xuesong Li, Binbin Wang, Dawei Yan, Hongjun Chen, Qinfang Liu and Zejun Li, "Ducks induce rapid and robust antibody responses than chickens at early time after intravenous infection with H9N2 avian influenza virus," *Virology Journal,* Last modified on April 11, 2019, Accessed March 30, 2021, https://virologyj.biomedcentral.com/articles/10.1186/s12985-019-1150-8.

[21] https://www.tyrantfarms.com/a-fowl-battle-ducks-vs-chickens/

[22] https://www.tyrantfarms.com/a-fowl-battle-ducks-vs-chickens/

[23] https://www.tyrantfarms.com/a-fowl-battle-ducks-vs-chickens/

[24] Sarah Moore, "Which Fowl Eat Mosquitoes?" *Pets on Mom.com,* Accessed April 26, 2021, https://animals.mom.com/fowl-eat-mosquitoes-11665.html.

adult mosquitoes.[25] Ducks eat flies and fly larvae,[26] helping keep your insect levels under control. They also eat bugs, worms, and all kinds of pests. Some varieties of ducks have been known to eat "weeds, snails, insects, and even small reptiles."[27] What a wonderful way to save money on feed bills and add some protein to your birds' diets!

If you want a sweet backyard pet... buying ducks may be an excellent option for you. Even though my ducks never became excellent pets, many have had the opposite experience. Ducks which are hatched at home and handled since birth can become very bonded to humans. In addition, certain breeds of ducks are very tame and friendly. Later in the book, we'll discuss varieties of ducks that make excellent pets.

Backyard poultry can provide one of the sweetest experiences of animal friendship that you may experience. There's something very endearing about the birds' total dependence on the humans. To have a fluffy duckling snuggled close to you, eyes closed contentedly, provides a profound sense of satisfaction. One of my friends has a disabled duck that lives with her in her home. She changes its diaper throughout the day, bathes it in her bathtub every morning, and procures a nanny for it if she wants to go on vacation. The duck plays with toys, watches TV, and knows how to roll a ball back and forth with its beak. The duck's owner even takes little Pippen on compassion visits to friends and family of people on hospice. One lady's last wish was to play with a duck. Since she was too weak to walk to the backyard and see a "normal" duck, Pippen's owner brought her to see this lady. What a wonderful last memory.

---

[25] Moore

[26] Mary, "How to Get Rid of Flies in Your Chicken Coop, Naturally," *Life is Just Ducky,* Accessed April 26, 2021, https://www.lifeisjustducky.com/how-to-get-rid-of-flies/.

[27] Samantha Johnson, "6 Duck Breeds to Raise for Eggs," *Hobby Farms,* Last modified March 15, 2019, Accessed on April 26, 2021, https://www.hobbyfarms.com/6-duck-breeds-to-raise-for-eggs-4/.

Pippen snuggles with a heart transplant survivor as she faces multiple health complications. She passed away three days later.

So, there's the scoop: the good, the bad, and the ugly. If you've read this far, I suspect you're one of those devoted duck fans who would do anything for your pets. Like me, you will stick with your ducks through thick and thin. You'll educate yourself, make a good choice, and enjoy your healthy, happy ducks for many years. It's time to jump in and buy those ducks! We think you'll never regret it.

Chapter 3: Bringing Your Duck Home: Which Breed to Buy?

Congratulations, you're about to become a duck owner! As you prepare to bring your precious ducklings home, breed is one of the most important things you need to take into consideration.

It's critical to clarify your goals and expectations for keeping ducks. Do you want a pet? Are you hoping for eggs? Are you looking for a meat source? Do you want to keep your neighborhood quiet by finding a less-vocal breed? Are you looking for beauty and elegance?

First, let's look briefly at the "best ducks" in each of these categories. Then, we'll take a more detailed look at all the duck breeds.

Best Breed of Ducks to Keep as Pets

If you are hoping for a pet duck, you'll need to select a breed known for its calm temperament and sweet nature. The owners of Tyrant Farms have done some research into the best breeds of ducks to keep as pets.[28] According to Aaron and Susan, Saxony ducks tie for first place. With their gentle personality and lack of aggression, they make perfect pets. In fact, they are common pets in Australia.[29]

Right at the top with the Saxony are the Muscovy ducks. This comical-looking breed has a great reputation on duck social media groups. Following the Saxony and the Muscovy are the Silver Appleyard, Welsh Harlequin, and Black Swedish duck.[30]

If you're looking for a loveable pet for your child, look no further!

Best Breed of Ducks for Eggs

The Khaki Campbell wins the prize for the best duck for egg production. Campbells can "lay as many as 340 eggs per year!"[31] That's up to one hundred forty more eggs per year

---

[28] Aaron and Susan, "How to Get Your Ducks to Like You: Three Tips," *Tyrant Farms,* Last modified on October 6, 2017, Accessed March 30, 2021, https://www.tyrantfarms.com/how-to-get-your-ducks-to-like-you-three-tips/.

[29] "Saxony Duck," Burke's Backyard, Accessed March 30, 2021, https://www.burkesbackyard.com.au/fact-sheets/pets/pet-road-tests/saxony-duck/.

[30] "How to Get Your Ducks to Like You"

[31] Samantha Johnson, "6 Duck Breeds to Raise for Eggs," *Hoppy Farms,* Last modified March 15, 2019, Accessed on April 26, 2021, https://www.hobbyfarms.com/6-duck-breeds-to-raise-for-eggs-4/.

than the Pekin duck, the most common farmyard duck.[32] Other breeds of ducks that are excellent egg layers include Indian Runners, Buff, Welsh Harlequin, Magpie, and Ancona.[33] When you choose an egg-layer, you will be treated with wonderful, mild, large, creamy eggs at your breakfast table every morning!

Best Breed of Ducks for Meat

When looking for a meat breed, you will want to choose a duck variety that has a higher meat volume, smaller percentage of fat, and better flavor.[34] To meet these qualifications, your best bet is the Pekin. This common white barnyard duck produces delicious, mild meat.[35] A large percentage of Pekin meat is luscious dark meat that we all love to find in our chickens, ducks, and turkeys.[36] Other meat ducks include Moulard, Muscovy, Aylesbury, and Rouen.

Quietest Duck Breed

All duck owners agree that Muscovy ducks are the quietest breed of duck. When first-time duck owners post in concern over their noisy pets, seasoned owners recommend, "Buy Muscovies." Muscovies are an odd-looking kind of duck. They do not spend as much time in water and perch on roosts like chickens. Most of the time, Muscovy ducks hardly make a sound.[37]

If you are looking for a more traditional duck that is still quiet, try the Khaki Campbell duck, Cayuga, Magpie, or Saxony.[38] Runner, Crested, Swedish, and Appleyard ducks are often quiet as well.[39]

An Overview of Duck Breeds

Now, we will look a little more closely at the individual breeds of ducks.

[32] "Pekin Duck Breed: Everything You Need to Know," *Happy Chicken,* Last modified November 3, 2020, Accessed April 26, 2021, https://www.thehappychickencoop.com/pekin-duck-breed-everything-you-need-to-know/.

[33] "6 Duck Breeds to Raise for Eggs"

[34] "Best Meat Duck Breeds," *The Happy Chicken Coop,* Last modified on November 24, 2020, Accessed April 26, 2021, https://www.thehappychickencoop.com/best-meat-duck-breeds/.

[35] "Best Meat Duck Breeds"

[36] "Best Meat Duck Breeds"

[37] April Lee, "9 Quietest Duck Breeds," Accessed April 26, 2021, https://farmhouseguide.com/quietest-duck-breeds/.

[38] https://farmhouseguide.com/quietest-duck-breeds/

[39] https://farmhouseguide.com/quietest-duck-breeds/

This section of the book provides a thorough encyclopedia of duck breeds. Feel free to skim through it at your leisure to select the duck that is right for you. If you've already decided the breed of your duck, you are welcome to skip this section and move to practical steps to raising ducks.

White Pekin

The White Pekin is a large, white, domesticated duck with an orange bill and feet. Pekins lay large, pale, pearly white or gray eggs. Pekins are by far the most popular and well-known breed of farmyard duck.[40] As we discussed above, Pekins are a top pick for meat, and they are also excellent egg layers. This makes them a top pick for commercial farms and backyard hobby farms. The Cape Coop explains that "Around 90% of the duck meat produced in America comes from Pekins."[41]

---

[40] "Great Backyard Duck Breeds," Accessed April 26, 2021, *The Cape Coop*, https://thecapecoop.com/great-backyard-duck-breeds/.
[41] "Great Backyard Duck Breeds," https://thecapecoop.com/great-backyard-duck-breeds/.

When you imagine "ducks on a farm," what image comes to mind? Most likely, you immediately imagine a flock of white ducks and geese. If so, you're thinking of a Pekin.

For this reason, Pekins are also the most common duck that is sold at Tractor Supply Companies. During "chick days," little Pekin ducklings can be seen peeping alongside the other chicks and ducklings at the store. Fascinated by their cute peeps, people find them irresistible and take a few home.

This is the reason you will often see these large, white ducks swimming in city ponds or begging for bread in the park. Mostly, these big white pond birds are rejects that have been cast aside by naïve owners. Backyard farmers who were not prepared for the work and investment involved in raising ducks later took them to the lake and let them go.

Questions for consideration:

1. Are you looking for a strong, sturdy breed of ducks for egg production?
2. Do you live in an area far from neighbors who might complain about the noise and smell that Pekins can create?
3. Are you committed for the long haul?
4. Are you looking for a great duck for meat production?

If your answer is "yes," Pekins might be the breed for you.

Black East Indian

The Black East Indian duck is a tall, upright black duck with a greenish tinge to its feathers. They are a bantam variety of duck.[42] This means they are small in size. The purpose of raising Black East Indian is typically for looks. Paul Ives says of the Black East Indian duck, "In 1943 the committee of three professional artists invited to select the most beautiful bird in the Boston Poultry Show, from a purely artistic standpoint . . . selected a Black East Indian drake as the most beautiful bird among 5000 specimens of all varieties of land and waterfowl."[43] They make good mothers and take good care of eggs.[44] They can fly, so make sure you keep them in a cage with a roof.[45]

Questions to consider:

1. Are you looking for an ornamental breed for shows or for looks?

---

[42] "Black East Indian Ducks," *Poultry Keeper,* Accessed April 26, 2021, https://poultrykeeper.com/duck-breeds/black-east-indian-ducks/.
[43] "Bantam Ducks Make Good Pets," *Ashton Waterfowl,* Accessed April 26, 2021, https://ashtonwaterfowl.net/bantam_ducks.htm.
[44] "Bantam Ducks Make Good Pets," *Ashton Waterfowl,* Accessed April 26, 2021, https://ashtonwaterfowl.net/bantam_ducks.htm.
[45] "Black East Indian Ducks," https://poultrykeeper.com/duck-breeds/black-east-indian-ducks/.

2. Do you need a reliable mother who can incubate duck eggs, whether for business or for pleasure?
3. Are you seeking a small duck that doesn't take as much space or consume as much food?
4. Do you have a secure enclosure that can ensure that these ducks do not fly away?

If so, then a Black East Indian duck might be a great option for you!

Blue Swedish

With a soft, gray-blue back, dark head, and white neck, the Blue Swedish duck is a beautiful breed to add to your flock. The duck originated in current-day Germany and Poland in the 1800s.[46] These ducks can weather a variety of conditions and are generally considered healthy and hardy.[47] Blue Swedish ducks make good pets because of their sweet personality.[48] They are good backyard ducks since they are less likely to fly out of the yard.[49]

Sometimes, they lay eggs that match the blue or gray tint of their feathers, though generally their eggs are white.[50]

Questions to consider:

1. Do you need a long-lasting duck who will live for many years and perform in all kinds of weather?
2. Are you looking for a pet for your children and grandchildren?
3. Do you need a backyard duck who cannot fly?
4. Are looks important to you?

If so, a Blue Swedish may be your duck of choice.

Silver Appleyard

The Silver Appleyard is a full-size duck with beautiful golden, tan, brown, and gray markings. The males have dark heads, like a mallard, with a white ring around the neck.

---

[46] "Swedish Blue," *Wikipedia,* Accessed April 26, 2021, https://en.wikipedia.org/wiki/Swedish_Blue.
[47] "Blue Swedish Ducks (Anas platyrhynchos domesticus)," *Beauty of Birds,* Accessed April 26, 2021, https://www.beautyofbirds.com/blueswedishducks.html.
[48] "Blue Swedish Ducks (Anas platyrhynchos domesticus)," https://www.beautyofbirds.com/blueswedishducks.html.
[49] "Blue Swedish Ducks (Anas platyrhynchos domesticus)," https://www.beautyofbirds.com/blueswedishducks.html.
[50] "Blue Swedish Duck," *McMurray Hatchery,* Accessed April 26, 2021, https://www.mcmurrayhatchery.com/blue_swedish.html.

They also have a dark rump, white wings with black tips, and brown markings on the chest and back. Female Appleyard ducks have tan necks with dappled gray and brown on the head and back.

Reginald Appleyard created this breed in the 1930s.[51] His goal was to develop a beautiful breed with large and good meat.[52] The Silver Appleyard breed has been called "A Great All-Round Duck."[53] The Appleyard duck lays a moderate number of eggs yearly (200-270).[54]

The full-size Appleyard Duck is a good option for pets, meat, and eggs.[55] They are also great show animals! These ducks don't fly, so it's a great option for keeping in the backyard.[56] The Appleyard duck has received rave reviews from duck owners. Listen to a few of these comments: "Five stars! Amazing; the BEST. They are the ultimate duck—there is no better."[57]

Another five-star reviewer said, "I have kept many breeds of duck, but to date I have found the Silver Appleyards to be one of the best. The egg laying capabilities are second only to Khaki Campbells, and the meat quality is also excellent."[58]

Another added, "A very good all-round duck. A very attractive bird; gentle, yet sturdy and strong. If brought up in the proper environment, they are a truly rewarding pet and a real looker if showing is your thing. Highly recommended!"[59]

Questions to consider:

1. Are you hoping to find an absolutely gorgeous duck?
2. Are you looking for excellent egg production?

---

[51] "The Silver Appleyard: A Great All-Round Duck," *The Modern Homestead,* Accessed April 26, 2021, https://www.themodernhomestead.us/article/Silver+Appleyard.html.

[52] "The Silver Appleyard: A Great All-Round Duck," https://www.themodernhomestead.us/article/Silver+Appleyard.html.

[53] "The Silver Appleyard: A Great All-Round Duck," https://www.themodernhomestead.us/article/Silver+Appleyard.html.

[54] "The Silver Appleyard: A Great All-Round Duck," https://www.themodernhomestead.us/article/Silver+Appleyard.html.

[55] Kim Irvine, "Appleyard Duck Breed – Everything You Need to Know," Last modified on November 26, 20218, *Domestic Animal Breeds,* Accessed April 26, 2021, https://domesticanimalbreeds.com/appleyard-duck-breed-everything-you-need-to-know/.

[56] "Appleyard Duck Breed – Everything You Need to Know," https://domesticanimalbreeds.com/appleyard-duck-breed-everything-you-need-to-know/.

[57] "Silver Appleyard Reviews," *Omlet,* Accessed April 26, 2021, https://www.omlet.us/breeds/ducks/appleyard/reviews.

[58] "Silver Appleyard Reviews," https://www.omlet.us/breeds/ducks/appleyard/reviews.

[59] "Silver Appleyard Reviews," https://www.omlet.us/breeds/ducks/appleyard/reviews.

3. Do you need your ducks to be able to provide meat as well?
4. In addition, do you want an excellent pet?
5. Are you aware that Silver Appleyard ducks are a little more difficult to find than other duck breeds?

If you are looking for an all-round excellent duck, look no further!

Miniature Appleyard

The Miniature Appleyard duck is a beautiful Bantam. These small ducks are white or pale yellow with lovely brown and tan mottled markings on their head, back, and wings. Some varieties have dappled gray and silver speckles on their head, back, and wings. Males have green heads with "a silver ring" around the neck,[60] like an elegant man going out with a silver necktie. Both males and females also feature blue on their wings.[61]

Named for the man who developed the original full-sized Silver Appleyard breed, the Bantam variety was actually created by Tom Bartlett.[62]

Appleyards are an excellent choice if you would like to raise baby ducklings. They are dedicated, loyal mothers who will sit on eggs with gusto.[63]

Questions to consider:

1. Are you seeking a small duck that doesn't require as much space or consume as much food?
2. Are you looking for an absolutely gorgeous duck to adorn your backyard?
3. Do you need a reliable mother who can incubate duck eggs to sell as a business, or simply to raise more ducklings for your own backyard?
4. Are you comfortable with paying a higher price and spending more time in research to find a breeder who sells this rare duck?

If you answered yes to these questions, a Silver Appleyard Bantam is your duck of choice!

Silver Bantam

The Silver Bantam duck has some similarities to the Silver Appleyard Bantam duck, but there are some marked differences. Silver Bantams have tan heads and dark beaks, and less speckles on the head and back. Male Silver Bantams have greenish black on their

---

[60] "Silver Appleyard Miniature," *Omlet UK,* Accessed April 26, 2021, https://www.omlet.co.uk/breeds/ducks/silver_appleyard_miniature.
[61] "Silver Appleyard Miniature," https://www.omlet.co.uk/breeds/ducks/silver_appleyard_miniature.
[62] "Silver Appleyard Miniature," https://www.omlet.co.uk/breeds/ducks/silver_appleyard_miniature.
[63] "Silver Appleyard Miniature," https://www.omlet.co.uk/breeds/ducks/silver_appleyard_miniature.

heads and rumps, and brown speckles on their chests and wings. Both have a black tip on their bills.

Unlike the Silver Appleyard that bears his name, The Silver Bantam is the miniature duck breed that was actually bred by Mr. Appleyard himself.[64]

When ducklings are only eight weeks old, you can already tell the gender of the duck by its bill. Females have dark gray beaks, while males have the typical olive color of a mallard.[65]

Silver Bantams can fly, so be careful to fence them adequately and protect them from predators.[66]

Questions to consider:

1. Are you seeking a small duck that doesn't need as much space or consume as much food?
2. Do you have appropriate fencing to keep your flighty birds contained?

If so, the Silver Bantam may be a perfect choice for your backyard farm.

Call Duck

The Call duck is a small, white duck with a short orange bill. Some varieties of Call ducks are also tan or brown.

The Call duck is one of the sweetest, cutest types of ducks you can imagine. If you've seen a super cute video featuring a tiny white duck dressed up in a Christmas outfit or snuggled in a flowerpot, it's likely to have been a Call duck. This tiny duck is absolutely adorable.

Call ducks are easy to tame and make wonderful pets.[67]

The Call duck was named for its ability to make a high-pitched call that would attract wild ducks for hunters.[68] When considering purchasing a Call duck, make sure that you and your neighbors are prepared for the noise and chatter that Call ducks constantly engage in.[69]

---

[64] "Silver Appleyard Miniature," https://www.omlet.co.uk/breeds/ducks/silver_appleyard_miniature.
[65] "Silver Bantam Duck," *Raising Ducks,* Accessed April 26, 2021, https://afowlshome.com/types-of-fowl/ducks/domestic-ducks/bantam-ducks/silver-bantam-ducks/.
[66] "Silver Bantam Duck," https://afowlshome.com/types-of-fowl/ducks/domestic-ducks/bantam-ducks/silver-bantam-ducks/.
[67] "Call Duck," *Omlet,* Accessed March 30, 2021, https://www.omlet.us/breeds/ducks/call_duck/.
[68] "Call Ducks," *Wikipedia,* Accessed April 26, 2021, https://en.wikipedia.org/wiki/Call_duck.
[69] "Call Ducks," https://www.omlet.us/breeds/ducks/call_duck/.

Questions to consider:

1. Are you seeking a small duck that doesn't require as much space or consume as much food?
2. Are you looking for a gentle, tame animal who can become an excellent pet for your children and grandchildren?
3. Do you have a desire to dress up your ducks?
4. Do you live far from neighbors who would be bothered by the loud and high-pitched quacking that Call ducks are famous for?
5. Are you committed to searching for a breeder who sells Call ducks, even though the breed is not as common as other breeds?

If you answered "yes" to each of these questions, it's time to find yourself a Call duck!

Mallard

Many of us are familiar with the beautiful mallard. With their lovely iridescent blue wing patch, white accents, tan body, and brown chest, the males sport a white neck ring, pale olive yellow beak, and green head. Females are brown with mottled dark brown patterns, a distinctive brown eye stripe and white wing accents. Mallards are commonly seen in the wild throughout North America, Asia, and Europe.[70] Additionally, Mallards live in Africa and have spread to "New Zealand, Australia, Peru, Brazil, Uruguay, Argentina, Chile, Colombia, the Falkland Islands, and South Africa."[71] Eggs range from pale cream to tan and green.[72]

We've all seen wild ducks and geese flying overhead, so it should not come as a surprise that Mallard ducks can fly. When you keep them as pets, be aware that they could fly away if you do not keep them well confined.[73] In addition, Mallards do not lay very well, and they are noisy and full of energy.

Questions to consider:

1. Are you prepared to answer any concerns about the origins of your ducks, confirming that you did not capture your Mallards from the wild?
2. Do you live in an area far from neighbors who might complain about the noise and smell that Mallards can create?
3. Do you have a secure enclosure that can ensure these ducks do not fly away?

---

[70] "Mallards," *National Geographic,* Accessed April 26, 2021,
https://www.nationalgeographic.com/animals/birds/facts/mallard#:~:text=Mallard.
[71] "Mallard," *Wikipedia,* Accessed April 26, 2021, https://en.wikipedia.org/wiki/Mallard.
[72] "Mallard," https://en.wikipedia.org/wiki/Mallard.
[73] "Great Backyard Duck Breeds," https://thecapecoop.com/great-backyard-duck-breeds/.

If so, mallards are the duck for you!

Bali

The Bali is one of the most comical ducks to look at. Tall and upright, it has a long neck and a puff of feathers on its head. Its very long, straight appearance gives it a quizzical look that sparks the imagination of many duck dress makers. Bali ducks are often dressed up with cute spring hats, colorful dresses, suits, ties, and other outfits. The sight of a group of Bali ducks, dressed to the T, strutting down the street alongside humans, will bring a smile to any face.

Every year, Australia has a fashion show for ducks.[74] Over nine hundred thousand visitors each year come to see these ducks in their fancy outfits.[75] There are white ducks in their little white and blue hats, adorned with blue ribbons, paired with a puffy white dress with blue calico flowers and lace. Or you'll spot an elegant, shiny black duck with bright yellow daffodils on its vests. Gold and purple dresses with red frills, or bright pink dresses with checkered black and white edges. Their long necks might be adorned with red or purple satin or stiff, buttoned up brown suits. A duck might even be dressed in a fancy white wedding dress, complete with a hat, flowers, lace, and a veil. In short, their elegant outfits fit into several categories: "daywear, evening attire, bridal outfits... costumes from the 1800's era as well as favorites from the past."[76]

Is a fashion show the only reason to keep Bali ducks? No, the Bali duck is a great option to keep in the backyard. The Bali lays eggs, forages, and is very practical for backyard duck owners.[77] In fact, the duck is still used in Bali to eat harmful insects in rice paddies.[78] These ducks work hard in the mud and would be mortified to see their American and Australian counterparts prancing around in expensive outfits.

---

[74] Sara Barnes, "Stylish Ducks Waddle Down the Catwalk in Annual Fashion Show," Last modified August 20, 2014, *My Modern Met,* Accessed April 26, 2021, https://mymodernmet.com/australian-pied-piper-duck-show/.

[75] "Stylish Ducks Waddle Down the Catwalk in Annual Fashion Show," https://mymodernmet.com/australian-pied-piper-duck-show/.

[76] "Stylish Ducks Waddle Down the Catwalk in Annual Fashion Show," https://mymodernmet.com/australian-pied-piper-duck-show/.

[77] "Bali," *Omlet,* Accessed April 26, 2021, https://www.omlet.us/breeds/ducks/bali/.

[78] "Bali," https://www.omlet.us/breeds/ducks/bali/.

Bali ducks have been around for a long time. There are pictures of Bali ducks carved into rock in Bali.[79] Other names for the Bali duck are "Balinese Crested Duck" and "Crested Runner Duck."[80] Its eggs are blue and green.[81]

Questions to consider:

1. Do you enjoy duck shows and fashion?
2. Are you a seamstress who thrives on creating amazing outfits for poultry?
3. Do you need an excellent forager to get rid of pests and harmful insects?
4. Do you need a practical source of eggs?

If so, you may want to choose a Bali duck or an Indian Runner, which we will explore next.

Indian Runner

Indian Runners seem to take the cake for the funniest duck breed. The Modern Homestead says it like this. "I've always found waterfowl in general great fun to raise, so sheer enjoyment was my first criterion. One of the most fun breeds I've encountered are Runner ducks. For a couple of years, my 'poultry buddy' Mike next door had a flock of seven, and I loved to go over and watch their antics. They were the clowns of the barnyard, with their hyperactivity and their odd vertical stance—a busy troop of animated bowling pins, or caricatures of soldiers moving ramrod-straight with military precision. Like some flocks of wild birds, they all moved together, veering in one direction or another in absolute lockstep. As Mike put it, 'They move like they're one tissue.'"[82]

Maybe that's one reason that Indian Runner ducks also compete in the Australian Fashion Shows. Like Bali Ducks, they stand upright with a comical, humanoid look. Indian Runners can be white, tan, or gray. Other colors include "black, white, chocolate, blue, fawn, mallard, white, and trout."[83] There are many variations of Runner ducks, including Penciled Runner ducks, black, chocolate, Cumberland blue, blue dusky and apricot dusky, mallard runners, silver runners, and apricot trout and blue trout.[84] Some have dark gray heads, white necks, golden torsos, and white abdomens. Even when they're wearing their natural feathers, they look like they're wearing a suit!

---

[79] "Bali," https://www.omlet.us/breeds/ducks/bali/.

[80] "Bali Ducks," *Poultry Keeper,* Accessed April 26, 2021, https://poultrykeeper.com/duck-breeds/bali-ducks/.

[81] "Bali Ducks," https://poultrykeeper.com/duck-breeds/bali-ducks/.

[82] https://www.themodernhomestead.us/article/Silver+Appleyard.html.

[83] "Indian Runner Duck," *Oregon Zoo,* Accessed April 26, 2021, https://www.oregonzoo.org/discover/animals/indian-runner-duck.

[84] "Standard Colours of the Indian Runner Duck," *Indian Runner Duck Club,* Accessed April 26, 2021, https://www.runnerduck.net/standard-colours.php.

Most ducks are well known for their waddle, but Runner ducks don't waddle. You guessed it—they run![85] Other ducks build nests and incubate eggs, but Runner Ducks more often just lay their eggs on the go.[86] As they walk and run around, their eggs drop out on the ground![87] If you keep Indian Runner ducks for eggs, you'll be doing an Easter Egg Hunt every day!

Runner ducks make less noise, eat less processed food, incubate eggs less often, and swim less than other duck species.[88]

Questions to consider:

1. Do you live close to other neighbors and need a quieter duck breed?
2. Are you trying to avoid the excessive water mess that comes along with most breeds?
3. Are you interested in duck shows and fashion?
4. Are you looking for a comical backyard pet?

If your answer to these questions was yes, consider buying yourself a few Indian Runners for your backyard comedy show.

Aylesbury

The Aylesbury duck is white like a Pekin, but it has a paler bill. Originating in Aylesbury, England, the duck was initially raised for its soft down feathers.[89] Many English children drifted off to sleep snuggled under warm down comforters stuffed by Aylesbury duck feathers. The city was full of breeders who would sell the ducks to Londoners.[90] These ducks are large ducks, with large bodies.[91] This breed is more popular in England, but in the United States has been replaced by the Pekin. The breed can be raised for other purposes, but is mostly used for show.[92]

Questions to consider:

1. Are you interested in duck shows and looks?
2. Are you prepared to house a large, full-bodied duck?

---

[85] "Indian Runner Duck," *Wikipedia,* Accessed April 26, 2021,
https://en.wikipedia.org/wiki/Indian_Runner_duck.
[86] https://en.wikipedia.org/wiki/Indian_Runner_duck
[87] https://en.wikipedia.org/wiki/Indian_Runner_duck
[88] https://en.wikipedia.org/wiki/Indian_Runner_duck
[89] "Aylesbury Duck," *Wikipedia,* Accessed April 26, 2021, https://en.wikipedia.org/wiki/Aylesbury_duck.
[90] "Aylesbury Duck," *Wikipedia,* Accessed April 26, 2021, https://en.wikipedia.org/wiki/Aylesbury_duck.
[91] "Aylesbury Duck," *Heritage Poultry,* Accessed April 26, 2021,
https://heritagepoultry.org/blog/aylesbury-duck.
[92] "Aylesbury Duck," https://heritagepoultry.org/blog/aylesbury-duck.

If so, you may choose to purchase an Aylesbury duck.

Abacot Ranger

At first glance, the Abacot Ranger resembles the Silver Appleyard duck or Silver Bantam. The males have dark heads, dark rumps, white feathers, a white neck ring, and brown markings on the chest and wings. Females have tan to golden heads with predominantly white bodies and brown markings.

Another English duck, the Abacot Ranger has gained new names in the different countries it has been imported to.[93] In America it is the Hooded Ranger, in Germany and East Europe it is the Streicherente, and in France it is the Le Canard Streicher.[94]

Abacot Rangers lay a moderate number of eggs and have a friendly disposition.[95] This duck can't fly, so it's a great option for keeping in the backyard.[96] Like Silver Bantam ducks, their gender can be determined by their bill color. Females have dark gray beaks, while makes have the typical olive color of a mallard.[97]

Questions to consider:

1. Do you need a friendly pet for your backyard?
2. Do you need a trustworthy source of fresh duck eggs?
3. Is it important to you to know the gender from the start?
4. Do you need a duck breed that can't fly so you can more easily allow it to forage in your backyard without risking flying away?

If so, consider adding an Abacot Ranger duck to your outdoor menagerie.

Khaki Campbell Duck

The Khaki Campbell duck is a beautiful brown duck. Its plumage is uniformly light brown, the color of mahogany. The males' coloration varies between lighter tan and deeper copper, while the females are lighter in color.[98]

---

[93] "Abacot Ranger Ducks," *Poultry Keeper,* Accessed April 26, 2021, https://poultrykeeper.com/duck-breeds/abacot-ranger-ducks/.

[94] "Abacot Ranger Ducks," https://poultrykeeper.com/duck-breeds/abacot-ranger-ducks/.

[95] https://poultrykeeper.com/duck-breeds/abacot-ranger-ducks/

[96] https://poultrykeeper.com/duck-breeds/abacot-ranger-ducks/

[97] "Abacot Ranger Duck," *Raising Ducks,* Accessed April 26, 2021, https://www.raising-ducks.com/duck-breed-guide/abacot-ranger-duck/.

[98] "Khaki Campbell Duck: Everything You Need To Know," Last modified on October 27, 2020, Accessed April 26, 2021, https://www.thehappychickencoop.com/khaki-campbell-duck-everything-you-need-to-know/.

The Khaki Campbell duck is another duck that originated in England. Like the Silver Appleyard duck, it is named after the breeder that first created this variety of duck: Mrs. Adele Campbell.[99]

Khaki Campbells are multipurpose ducks. They make good mothers and can incubate eggs; they are moderately good layers; and they yield delicious meat.[100]

The breed can survive a wide range of temperature, from subzero temps to Saharan heat.[101] The breed is not prone to flying, making them ideal for keeping in the backyard.[102] Furthermore, they are not as noisy as many other duck breeds, which makes them a good choice for city life.[103]

Questions to consider:

1. Do you live near other neighbors who might be bothered by louder breeds of ducks?
2. Are you looking for a quieter breed?
3. Do you need a good source of eggs?
4. Are you looking for a duck who can provide excellent meat for consumption?
5. Do you need a hardy breed that can withstand all types of temperatures?
6. Do you need a flightless bird who can graze in the backyard without fear of flying away?
7. Do you need a reliable mother who can incubate duck eggs, whether for business or for pleasure?

If so, Khaki Campbells will be an excellent option for your farm or garden.

Cayuga

The Cayuga is a darker breed of duck with a black, green, and blue sheen. Like a mallard, it has a blue wing patch.

---

[99] "Khaki Campbell Duck," *Livestock Conservancy,* Accessed April 26, 2021,
https://livestockconservancy.org/index.php/heritage/internal/campbell.
[100] "Khaki Campbell Duck: Everything You Need To Know," *The Happy Chicken Coop,*
https://www.thehappychickencoop.com/khaki-campbell-duck-everything-you-need-to-know/.
[101] "Khaki Campbell Duck: Everything You Need To Know,"
https://www.thehappychickencoop.com/khaki-campbell-duck-everything-you-need-to-know/.
[102] "Khaki Campbell Duck: Everything You Need To Know,"
https://www.thehappychickencoop.com/khaki-campbell-duck-everything-you-need-to-know/.
[103] "Khaki Campbell Duck: Everything You Need To Know,"
https://www.thehappychickencoop.com/khaki-campbell-duck-everything-you-need-to-know/.

"Mr. R. Teebay of Fulwood, Preston, Lancashire, UK" believes the Cayuga descended from an English breed of black duck.[104]

Cayugas are easily tamed[105] and do not quack as loudly as other ducks,[106] making them great options for backyard farming. Backyard Poultry shares an interesting fact about their eggs. At the beginning of the laying season, their eggs are black, similar to the color of the adult birds. However, as the season progresses, the eggs become light gray, then blue, then green, and finally white![107]

Questions to consider:

1. Do you live near other neighbors who might be bothered by louder breeds of ducks?
2. Are you looking for a quieter breed?
3. Are you looking for a sweet, calm duck that will become an excellent pet for your children and grandchildren?

If so, consider buying a beautiful Cayuga duck.

Crested

Like the Bali duck, the Crested duck has a cheery puff of feathers on top of its head. However, unlike the Bali duck, these birds have a more typical duck-like posture. Most Crested ducks are white, but some can be "buff, blue, and gray (mallard coloration)."[108]

Crested ducks have likely existed for many years. Ducks with feathers on their heads have been seen in drawings that are two thousand years old.[109]

[104] "Cayuga Duck," *The Livestock Conservancy,* Accessed April 26, 2021, https://livestockconservancy.org/index.php/heritage/internal/cayuga.
[105] "Breed Profile: Cayuga Duck," *Backyard Poultry,* Accessed April 26, 2021, https://backyardpoultry.iamcountryside.com/poultry-101/cayuga-duck-breed-spotlight/.
[106] "The Cayuga Duck Originates From New York, Named After Lake Cayuga," *Wide Open Pets,* Accessed April 26, 2021, https://www.wideopenpets.com/cayuga-duck/.
[107] "Breed Profile: Cayuga Duck," https://backyardpoultry.iamcountryside.com/poultry-101/cayuga-duck-breed-spotlight/
[108] "Crested Duck: Characteristics, Origin, Uses & Full Breed Information," *Roy's Farm,* Last modified March 11, 2021, Accessed on April 26, 2021, https://www.roysfarm.com/crested-duck/.
[109] "Crested (duck breed)," *Wikipedia,* Accessed April 26, 2021, https://en.wikipedia.org/wiki/Crested_(duck_breed).

Crested ducks do not fly, so they are ideal for keeping in the backyard. Crested ducks tolerate all weather and climates. They lay large, off-white eggs.[110] Their egg production is moderate, and their main purpose is as show birds or pets.[111]

Questions to consider:

1. Are you interested in duck shows?
2. Do you want a unique ornamental duck?
3. Do you need a flightless bird who can graze in the backyard without fear of flying away?

If so, consider buying a unique Crested duck for your backyard farmstead.

Dutch Hookbill

The first thing you'll noticed when you meet a Dutch Hookbill duck is its unusual beak. Like a toucan's beak, the bill curves toward the ground. The Dutch Hookbill duck is colored with a green head, white chest, gray and brown body, and white rump. Females have white wing tips with brown bodies, speckled with darker brown.

Dutch Hookbills are extremely rare. There are only two hundred fifty to four hundred ducks of these species in the world.[112] In 1980, a man named Hans van de Zaan worked to keep the Hook Bill duck from extinction.[113] He saved fifteen ducks that were still alive after the pollution of the water where they lived.[114] These ducks then bred and reproduced to the two hundred to four hundred ducks that are still alive today. These ducks are mostly show ducks.[115]

These ducks do not quack as loudly as Pekins and other varieties of ducks.[116] They lay blue eggs.[117]

Questions to consider:

---

[110] "Crested Duck: Characteristics, Origin, Uses & Full Breed Information," https://www.roysfarm.com/crested-duck/.

[111] Crested Duck: Characteristics, Origin, Uses & Full Breed Information

[112] "Dutch Hookbill: Important Things To Note About This Duck Breed," Last modified on December 10, 2020, Accessed April 26, 2021, https://iduckn.com/dutch-hookbill-duck/.

[113] "Dutch Hookbill: Important Things To Note About This Duck Breed"

[114] "Dutch Hookbill: Important Things To Note About This Duck Breed"

[115] "Dutch Hookbill: Important Things To Note About This Duck Breed"

[116] "Hook Bill Ducks," Poultry Keeper, Accessed April 26, 2021, https://poultrykeeper.com/duck-breeds/hook-bill-ducks/.

[117] https://poultrykeeper.com/duck-breeds/hook-bill-ducks/

1. Are you interested in showing your ducks?
2. Do you live near other neighbors who might be bothered by louder breeds of ducks?
3. Are you looking for a quieter breed?
4. Are you prepared to do the extra homework to find this unique and rare breed of duck? Are you able to pay any additional fees necessary to purchase this unique animal?

If so, Dutch Hookbills are your duck of choice!

Crested Miniature

What a sweet and amazing duck! These adorable and hilarious ducks include the beauty of a Call duck with the pretty hairdo of Crested ducks. Some Crested Miniature ducks are white, like a Call duck. Others are white with black spots. Still others have the coloration of a Mallard, with blue wings and brown speckles over their bodies. There are even brown speckles in the hairdo!

The feathers on the Crested Miniature's head sprout from a lump of fat.[118] Because the crest comes from a mutation, one quarter of all Crested ducklings die before hatching because of genetic problems.[119]

The Crested Miniature may look cute, but it is very energetic and constantly on the move.[120]

Questions to consider:

1. Are you prepared for a lively and energetic duck?
2. Are you aware of the high death rate among hatching eggs?
3. Do you want a beautiful, comical, and cute duck?

If so, try buying the Crested Miniature duck.

Magpie

---

[118] "Crested Miniature," *Omlet,* Accessed April 26, 2021,
https://www.omlet.us/breeds/ducks/crested_miniature/.
[119] "Crested Miniature Ducks," *Poultry Keeper,* Accessed April 26, 2021, https://poultrykeeper.com/duck-breeds/crested-miniature-ducks/.
[120] "Crested Miniature," *Omlet,* Accessed April 26, 2021,
https://www.omlet.us/breeds/ducks/crested_miniature/.

What an adorable black and white spotted duck! Most Magpie ducks have a black saddle on the back and a black hood on their heads. Others have speckles on the back and chest. The bird resembles the European Magpie.[121]

Magpie ducks can make good pets, good foragers, nice show birds, good layers, and excellent meat birds.[122] They are great all-purpose birds.

In addition, farmers use Magpie ducks for controlling liver flukes.[123] Liver flukes are a type of parasite that affect humans and animals.[124] They enter the digestive system of sheep and cows through watercress infested with liver fluke larvae.[125] Ducks usefully sift through water and water plants, finding and eating the liver fluke larvae before they cause problems.

Magpie ducks come in a variety of colors: "Black and white, blue and white, chocolate and white, dun and white."[126] The Magpie duck occasionally walks upright like an Indian Runner duck.[127]

Magpie ducks do not fly, but they startle easily. They might jump over a fence if you scare them.[128]

Questions to consider:

1. Do you need a bird who can eliminate harmful insects from your yard and garden?
2. Are you interested in showing your birds?
3. Do you need a reliable source of eggs?
4. Are you looking for excellent meat production?
5. Do you have a liver fluke infestation that you need to bring under control?

If so, search for a Magpie duck to complete your farmyard.

Orpington

---

[121] "Magpie Duck: Characteristics, Origin, Uses & Full Breed Information," *Roys Farms,* Accessed April 26, 2021, https://www.roysfarm.com/magpie-duck/.

[122] "Magpie Duck: Characteristics, Origin, Uses & Full Breed Information"

[123] "Magpie Duck: Characteristics, Origin, Uses & Full Breed Information"

[124] "Parasites—Liver Flukes," *Centers for Disease Control and Prevention,* Accessed April 26, 2021, https://www.cdc.gov/parasites/liver_flukes/index.html.

[125] https://www.cdc.gov/parasites/liver_flukes/index.html

[126] https://www.roysfarm.com/magpie-duck/

[127] "Magpie duck," *Wikipedia,* accessed April 26, 2021, https://en.wikipedia.org/wiki/Magpie_duck.

[128] "Magpie Duck," *The Livestock Conservancy,* Accessed April 26, 2021, https://livestockconservancy.org/index.php/heritage/internal/magpie.

On the outside, the Orpington duck resembles the Khaki Campbell duck. Like the Campbell, the Orpington duck is light brown to tan. Occasionally, Orpington ducks are highlighted with white. However, show birds should be uniformly tan, fawn, or buff.[129] In addition, some Orpington ducks have gray heads.

Orpingtons are common "pond ornaments," swimming gracefully around the pond. They rival Pekins for meat quality.[130] They are ready to butcher in eight to ten weeks.[131] Unlike Magpie ducks, Orpingtons do not startle easily.[132] Most ducks panic and start quacking at the slightest disturbance, so this quality is helpful for keeping ducks on a farm.[133]

Orpingtons go broody easily and take good care of their young.[134] They lay a moderate number of eggs.[135] Orpingtons do a great job with taking care of pests. They eat "tadpoles, young frogs, mosquitoes, small lizards, slugs, snails, wild greens, and little crustaceans."[136]

Orpingtons are also easy to tame.[137] They are smart and can be taught to stay within their boundaries.[138] Orpingtons make a great option for first-time backyard duck owners.

Questions to consider:

1. Are you looking for a calm duck that does not startle easily?
2. Do you need steady and stable additions to your backyard farm?
3. Do you want a beautiful duck to swim gracefully on your pond?
4. Do you need help with backyard pests and harmful insects?
5. Are you looking for an absolutely amazing meat bird that is ready to harvest in just a few weeks?
6. Do you need a gentle, loving pet for your children and grandchildren?
7. Are you looking for a smart duck that will not wander away when trained?
8. Are you looking for a good source of eggs?
9. Do you need a reliable mother who can incubate duck eggs, whether to sell in your business or to grow your backyard farm?

---

[129] "Orpington Duck," *Wikipedia,* Accessed April 26, 2021, https://en.wikipedia.org/wiki/Orpington_Duck.
[130] "Buff Orpington Duck Breed: Everything You Need To Know," *The Happy Chicken Coop,* Accessed April 26, 2021, https://www.thehappychickencoop.com/buff-orpington-duck-breed-everything-you-need-to-know/.
[131] "Buff Orpington Duck Breed: Everything You Need To Know"
[132] "Buff Orpington Duck Breed: Everything You Need To Know"
[133] "Buff Orpington Duck Breed: Everything You Need To Know"
[134] "Buff Orpington Duck Breed: Everything You Need To Know"
[135] "Buff Orpington Duck Breed: Everything You Need To Know"
[136] "Buff Orpington Duck Breed: Everything You Need To Know"
[137] "Buff Orpington Duck Breed: Everything You Need To Know"
[138] "Buff Orpington Duck Breed: Everything You Need To Know"

If so, an Orpington duck is a wonderful option!

Rouen

Crossing the large, heavyset body of the Pekin with the elegant colors of the Mallard creates the beautiful Rouen duck. The Rouen male has a greenish black head, white ring around the neck, olive yellow beak, gray underbelly, brown back, and a blue patch with white accents on the wing. The female is brown with dark brown accents on the back, chest, and wings. The female has a light brown stripe on her eye. The shiny blue patch on the female's wings is accented with beautiful white stripes on either side.

Unlike many duck breeds, which were developed in England, Rouens are a French duck.[139]

Rouens are primarily meat birds.[140] Their plump, stocky bodies can weigh up to ten pounds.[141] As meat birds, they do not lay many eggs per year.[142] The Happy Chicken Coop comments that Rouen meat "has a high fat percentage that floats to the surface and can be used to make some very delicious noodles or rendered for other culinary uses."[143]

These heavy birds are unlikely to fly away and are calm backyard companions.[144]

Questions to consider:

1. Are you looking for a calm duck to keep in your backyard?
2. Are you looking for an absolutely amazing meat duck that produces up to ten pounds of meat per bird?
3. Do you need a flightless bird who can graze in the backyard without fear of flying away?

If you answered yes to each of these questions, consider buying the reliable Rouen duck.

Rouen Clair

The Rouen Clair is another French duck related to the Rouen.[145] Rouen Clair females have lovely brown or fawn tones, black accents on feathers, a white stripe by the eye, and a

---

[139] "Rouen," *Omlet,* Accessed April 27, 2021, https://www.omlet.us/breeds/ducks/rouen/.

[140] "Rouen—Non-Industrial Duck," https://livestockconservancy.org/index.php/heritage/internal/rouen.

[141] https://livestockconservancy.org/index.php/heritage/internal/rouen

[142] "Rouen—Non-industrial Duck," Livestock Conservancy, Accessed April 27, 2021, https://livestockconservancy.org/index.php/heritage/internal/rouen.

[143] "Best Meat Duck Breeds," https://www.thehappychickencoop.com/best-meat-duck-breeds/.

[144] "Poultry Breeds—Rouen Duck," *Breeds of Livestock, Department of Animal Science,* Accessed April 27, 2021, http://afs.okstate.edu/breeds/poultry/ducks/rouen/index.html/.

[145] "Rouen Clair," *Omlet,* Accessed April 27, 2021, https://www.omlet.us/breeds/ducks/rouen_clair/.

beautiful blue patch on the wing. The white stripes above and below the eye, the white bib, and the blue and white accented wings make the Rouen Clair truly beautiful.

Rouen Clairs lay better than Rouens, but they are primarily meat birds.[146] Tending toward obesity, they are not good flyers and will stay in your yard.[147] Rouen Clairs also make good pets.[148] They need deep water for mating.[149]

Questions to consider:

1. Do you need a flightless bird who can graze in the backyard without fear of flying away?
2. Do you have sufficient water for mating?
3. Are you looking for a sweet, calm duck that will become an excellent pet for your children and grandchildren?

If so, the Rouen Clair is your duck of choice!

Welsh Harlequin

Welsh Harlequins are truly beautiful. The females are cream colored with slate gray beaks. Over their bodies, they are accented with silver, gold, and black. What silvery beauties!

Males look similar to Mallards but have more white and silvery accents on their bodies. Their chests are the color of copper, their backs and sides are gold, and their tails are black. Welsh Harlequin drakes have olive yellow beaks, green heads, and a white ring around the neck.

The Welsh Harlequin breed comes from a Khaki Campbell mutation.[150] Leslie Bonnet noticed the beautiful mutation and continued to successfully breed it.[151]

If you've been around ducks for any length of time, you know that some ducks have a very hard time with change. Furthermore, some ducks are easily scared, flighty, prone to panic, and afraid of humans.

---

[146] "Rouen Clair," *Omlet,* Accessed April 27, 2021, https://www.omlet.us/breeds/ducks/rouen_clair/.
[147] "Rouen Clair," *Omlet,* Accessed April 27, 2021, https://www.omlet.us/breeds/ducks/rouen_clair/.
[148] "Rouen Clair," *Omlet,* Accessed April 27, 2021, https://www.omlet.us/breeds/ducks/rouen_clair/.
[149] "Rouen Clair," *Omlet,* Accessed April 27, 2021, https://www.omlet.us/breeds/ducks/rouen_clair/.
[150] Henke, Jodi, "Raising Welsh Harlequin Ducks," *Successful Farming,* November 25, 2013, https://www.agriculture.com/family/living-the-country-life/raising-welsh-harlequin-ducks.
[151] "Welsh Harlequin Duck: Everything You Need to Know," *The Happy Chicken Coop,* Last modified on November 12, 2020, Accessed on April 27, 2021, https://www.thehappychickencoop.com/welsh-harlequin-duck-everything-you-need-to-know/.

Welsh Harlequins, however, are a different story. Compared to other duck breeds, Welsh Harlequins are not as easily frightened.[152] Furthermore, they are flexible and adaptable. They can adjust to "new surroundings and nearly any type of environmental conditions" without having a mental breakdown.[153] Further, Welsh Harlequins love spending time with humans and do not run away from them.[154] They are responsive to training and to coming to humans for attention and treats.[155] They are easy to put back in their coop in the evening.[156] Because of all these factors, Welsh Harlequins are excellent pets![157]

Harlequins lay a moderate number of eggs per year.[158] Harlequins "are excellent foragers and love to dine on bugs and pasture greens."[159]

If you are specifically looking for male or female Welsh Harlequins, you are in luck. Right after they hatch, female Welsh Harlequin ducklings have light beaks with a dark blotch on the end of the bill, while males have darker beaks.[160] This handy clue helps you take home the bird that will fit your flock best.

If you are looking for a first-time pet duck, a Harlequin might be an excellent, tame, and beautiful option for you!

Questions to consider:

1. Do you want to own an absolutely stunning duck whose feathers are accented in gold and silver?
2. Are you looking for a sweet, calm duck that will become an excellent pet for your children and grandchildren?
3. Do you need a flexible, adaptable duck that is not easily frightened by changes?
4. Are you looking for a smart duck that is easily trained?
5. Do you need help with backyard pests and harmful insects?
6. Do you need a moderate source of eggs?

---

[152] "Raising Welsh Harlequin Ducks"

[153] https://www.thehappychickencoop.com/welsh-harlequin-duck-everything-you-need-to-know/

[154] https://www.thehappychickencoop.com/welsh-harlequin-duck-everything-you-need-to-know/

[155] https://www.thehappychickencoop.com/welsh-harlequin-duck-everything-you-need-to-know/

[156] https://www.thehappychickencoop.com/welsh-harlequin-duck-everything-you-need-to-know/

[157] Jodi Henke, "Raising Welsh Harlequin Ducks," *Successful Farming,* Last modified on November 25, 2013, Accessed on April 27, 2021, https://www.agriculture.com/family/living-the-country-life/raising-welsh-harlequin-ducks.

[158] https://www.agriculture.com/family/living-the-country-life/raising-welsh-harlequin-ducks

[159] https://www.agriculture.com/family/living-the-country-life/raising-welsh-harlequin-ducks

[160] "Welsh Harlequin Duck," *Murray McMurray,* Accessed April 27, 2021, https://www.mcmurrayhatchery.com/welsh_harelquin_duck.html

If so, try purchasing a lovely Welsh Harlequin duck.

Black Swedish

Black Swedish ducks are mostly black, as their name suggests. However, they also have a white patch on the chest and neck, as well as white accents on the tail. Black Swedish ducks are colored this way from the time they hatch.

Black Swedish ducks are used as breeding stock for Blue Swedish ducks.[161]

These birds are not as outgoing and energetic as other duck breeds; they are reserved and peaceful.[162] Black Swedish eggs are usually white, but they can also be gray or blue.[163] Occasionally, they produce green eggs as well.[164]

Questions to consider:

1. Do you need a quiet, low-energy duck?
2. Are you looking for a beautiful, striking backyard companion?
3. Are you interested in showing off your unusually colored green, blue, white, and gray eggs?

If your answer to these questions is yes, consider purchasing a Black Swedish duck.

Muscovy

Have you ever heard of a duck that doesn't like to swim? How about a duck without webbed feet? Or a duck that flies and perches in trees? Or a duck that has a knobby red face? If you've never heard of this type of duck, you've never met a Muscovy.

Muscovy are a truly unique breed of duck. Whereas most domestic ducks have the scientific name Anas platyrhynchos domesticus, Muscovy ducks come from a different group all together. Their scientific name is Cairina moschata domestica.[165]

Shiny black feathers, iridescent green decorations, and white accents make the Muscovy breed a beautiful type of duck.

---

[161] "Black Swedish Ducklings," *Purely Poultry,* Accessed April 27, 2021, https://www.purelypoultry.com/black-swedish-ducklings-p-862.html.
[162] "Black Swedish Ducklings"
[163] "Black Swedish Duck," *Murray McMurray,* Accessed April 27, 2021, https://www.mcmurrayhatchery.com/black_swedish_duck.html.
[164] "Ducklings: Black Swedish," *My Pet Chicken,* Accessed April 27, 2021, https://www.mypetchicken.com/catalog/Waterfowl/Ducklings-Black-Swedish-p2565.aspx.
[165] "Mulard," *Wikipedia,* Accessed April 27, 2021, https://en.wikipedia.org/wiki/Mulard.

The most distinctive visual feature of a Muscovy duck is the collection of caruncles, or knobs of red tissue, that grow on their faces. The growths can sometimes be black, as well as red.[166] These unsightly growths make Muscovy ducks stand out from other ducks. For this reason, many backyard duck owners hesitate to buy Muscovy ducks. Some call them the "ugly duckling."

Muscovy ducks also have a crest on their heads. These feathers can lay flat, but they can raise this crest when they are excited, scared, or upset.[167]

Muscovy ducks can also be pure white with black wings and tail. At other times, they can be completely white. Some are gray, speckled, or gray with black and white patterns. Others are brown or shine almost completely green. The official Muscovy coloration possibilities include "Black, White, Blue, Chocolate, Bronze, and Lavender."[168]

Let's talk about some of the color combinations possible within the Muscovy breed:

- Chocolate. Chocolate ducks are almost universally brown, with occasional spots or speckles of white or black.
- Atipico Dusky. These birds are black, covered with a beautiful green sheen from head to tail.
- Chocolate Pied. With chocolate or brown colored back and head, this bird has a white bib, thick white eye stripe, and white chest and wings.
- Black. A truly black Muscovy has wings that are black on top and bottom.[169] Black birds may have white on their heads or backs.
- Blue Muscovy ducks have a lovely gray, blue, or silver hue. Their neck and head may be the color of rust.[170]
- Canizie. The word canizie means that the birds will have a white head.
- Black barred. A barred duck has mottled white and black (or white and brown) on the wings and back. However, the barred coloration only lasts until after the first several molts.[171]

---

[166] "Muscovy Duck: Eggs, Facts, Care Guide and More…" *The Happy Chicken Coop,* Accessed April 27, 2021, https://www.thehappychickencoop.com/muscovy-duck/.
[167] "Muscovy Duck: Eggs, Facts, Care Guide and More…" *The Happy Chicken Coop,* Accessed April 27, 2021, https://www.thehappychickencoop.com/muscovy-duck/.
[168] "What color is it?" *The Ugly Duck Farm,* Accessed April 27, 2021, http://muscovy.us/what_color_is_it.
[169] "What color is it?"
[170] "What color is it?"
[171] "What color is it?"

- Dark Ripple. A ripple coloration means that the barred or mottled appearance will last the duck's entire life, and will not disappear after the first several molts.[172]
- Blue Fawn. This breed has blue mixed with brown or fawn. It may have white coloration as well.
- Silver. The silver Muscovy is a lighter shade of gray.[173]
- Lavender. The lavender Muscovy is a darker shade of gray. You will notice a slightly lilac or lavender colored hue throughout the silver color.[174]
- Magpie. A magpie Muscovy looks just like a real magpie, with its black head, black tail and rump, and black shoulders and wing tops. Since magpie Muscovies can fly, they look even more like the real thing.
- Cream. Strangely and ironically, the cream Muscovy contains lavender and chocolate genes.[175] Even though cream Muscovies may look like they are white, they have some overtones of off white.
- Buff Cream. The buff cream breed has some of the chocolate showing through.
- White. White Muscovies are true, pure white, like the satin on a bridal gown.
- Silver Lavender. These birds are light blue gray in color.[176]

In the wild, the Muscovy breed lives in Central America, Mexico, and South America.[177] In the United States, groups of Muscovy ducks live on their own in the wild. Like wild horses which have escaped and are now living in the wild, Muscovies also live in certain states in the United States, such as Texas, Florida, Massachusetts, Louisiana, Hawaii, and Canada, as well as Australia, New Zealand, and Europe.[178]

There are many significant differences between Muscovy ducks and other domestic ducks.

First, Muscovy ducks make a different sound than Mallard derived ducks. Most ducks quack loudly and harshly, but Muscovy ducks make a distinctive, quieter call. The drake's call has been described as a "low breathy call," while the hen's call has been described as a "quiet trilling coo."[179] If you live in the city, this quality of Muscovy ducks will be a

---

[172] "What color is it?" *The Ugly Duck Farm,* Accessed April 27, 2021, http://muscovy.us/what_color_is_it.
[173] "What color is it?"
[174] "What color is it?"
[175] "What color is it?"
[176] "What color is it?"
[177] "Muscovy duck," *Wikipedia,* Accessed April 27, 2021, https://en.wikipedia.org/wiki/Muscovy_duck.
[178] https://en.wikipedia.org/wiki/Muscovy_duck
[179] https://en.wikipedia.org/wiki/Muscovy_duck

definite selling point. If you have been raising Pekins, you may be tired of their loud, grating screams. However, Muscovies will lull you to sleep with their gentle, dove-like cooing.

Most ducks waddle on their ample webbed feet, but Muscovy ducks are different. Muscovy duck are not "most ducks." They have claws and perch in trees. Your pet Muscovy duck will enjoy having a perch in their pen so they can roost at night.

Female Muscovy ducks can fly, so if you keep them as backyard pets, make sure they are in a safe enclosure.[180] Male ducks are hindered from flying by their weight.[181]

Muscovy ducks lay white or yellowish eggs.

Muscovy ducks grow to be about fifteen pounds, a giant duck.[182] Compare this to the Pekin duck, which only grows to be seven to nine pounds.[183] For this reason, Muscovy ducks are an excellent choice for meat birds. Here's what the Happy Chicken Coop has to say about a Muscovy's meat producing capacity: "It has the highest meat yield of any duck. The meat is 98% fat-free, is much less greasy than other ducks, and there is approximately 50% more breast meat than a standard duck. It also has fewer calories and fat than a turkey pound for pound."[184]

Muscovy ducks do not swim as much as mallard derived ducks.[185] They do enjoy water, however.

Muscovy ducks have a large, wide tail, like a bird of flight. The Happy Chicken Coop explains the following about Muscovy ducks' tails: "Muscovies are quiet, peaceful ducks with personality. They 'talk' with their tail, wagging it furiously when animated or happy, much like a dog does... If upset, happy, or excited, they wag their tails a great deal, and the males can also puff and hiss."[186] Would you like a pet duck that wags its tail, flies, and roosts in trees like a prehistoric dinosaur? If so, a Muscovy Duck may be a perfect choice for you.

Questions to consider:

1. Can you tolerate unusual looks in ducks?

[180] "Muscovy Duck: Eggs, Facts, Care Guide and More…"
[181] "Muscovy Duck: Eggs, Facts, Care Guide and More…"
[182] https://en.wikipedia.org/wiki/Muscovy_duck
[183] "Pekin Duck," *Columbian Park Zoo,* Accessed April 27, 2021,
http://www.lafayette.in.gov/DocumentCenter/View/1806/Pekin-Duck-
PDF#:~:text=Physical%20Characteristics%3A%20Weigh%207%2D9,Ducklings%20are%20yellow.
[184] https://www.thehappychickencoop.com/muscovy-duck/
[185] "Muscovy Duck: Eggs, Facts, Care Guide and More…"
[186] "Muscovy Duck: Eggs, Facts, Care Guide and More…"

2. Are you looking for a duck breed that doesn't need as much water?
3. Do you have a pen or enclosure that can safely secure a flighty female Muscovy duck?
4. Do you want a friendly duck that can become a pet for yourself or your kids and grandkids?
5. Do you live close to neighbors who might be bothered by constant loud quacking?
6. Are you looking for a quieter breed of duck?
7. Are you looking for a duck who can provide high quantities of excellent meat for consumption?
8. Do you need a duck that can produce low-fat breast meat?

If so, Muscovy ducks are the choice for you!

Moulard

Moulards are a unique cross between two genera of ducks: Muscovies (Cairina moschata domestica) and regular duck (Anas platyrhynchos domesticus).[187] Like mules, Moulards are unable to reproduce naturally.[188] Instead, they are produced by artificial insemination of female Pekin ducks with Muscovy drakes.[189] Moulards (also called Mulards) are filial 1 hybrids, or the offspring that result from the crossing of two different types of parents.[190]

Moulards are bred for their Foie Gras production. Foie gras is a delicacy made from the abnormally oversized fatty liver of a force-fed animal.[191] Foie Gras is eaten in France, and its "flavour is described as rich, buttery, and delicate, unlike that of an ordinary duck or goose liver."[192]

As far back as the ancient Egyptians, force feeding was used to produce foie gras.[193] Egyptian carvings show men sitting among rows and rows of ducks, holding their necks while forcing food into their open mouths. The same practice later took place in Greece, as the earliest reference to goose fattening was recorded by Cratinus.[194]

Birds harvested for foie gras are force-fed several times a day in a process called gavage.[195] Animal Equality describes the process: "Workers shove metal tubes down the birds'

---

[187] "Mulard," *Wikipedia,* Accessed April 27, 2021, https://en.wikipedia.org/wiki/Mulard.

[188] "Mulard," *Wikipedia,* Accessed April 27, 2021, https://en.wikipedia.org/wiki/Mulard.

[189] "Mulard," *Wikipedia,* Accessed April 27, 2021, https://en.wikipedia.org/wiki/Mulard.

[190] "F1 Hybrid," *Wikipedia,* Accessed April 27, 2021, https://en.wikipedia.org/wiki/F1_hybrid.

[191] "Foie Gras," *Wikipedia,* Accessed April 27, 2021, https://en.wikipedia.org/wiki/Foie_gras.

[192] https://en.wikipedia.org/wiki/Foie_gras

[193] https://en.wikipedia.org/wiki/Foie_gras

[194] https://en.wikipedia.org/wiki/Foie_gras

[195] https://en.wikipedia.org/wiki/Foie_gras

throats and pump their stomachs full of far more food than they would ever want to eat. This process, also known as 'gavage,' is repeated multiple times a day. The force feeding causes the birds' livers to swell up to ten times their natural size."[196]

In the past, geese were used for foie gras production.[197] Moulards present a clear advantage over geese because the ducks are smaller, cheaper to feed, calmer, and less likely to attack.[198] Furthermore, the cross between Muscovy and Pekin is ideal, since Moulards are more passive than Muscovies and less likely to resist the force feeding that is necessary to produce foie gras.[199]

Rescues like Farm Sanctuary exist to rescue Moulards from this destiny. Carrie is a beautiful Moulard duck who was rescued from a Foie Gras farm.[200] In your backyard, you can provide a loving home to these mistreated birds. You can give them a life of freedom, joy, and satisfaction.

Questions to consider:

1. Are you interested in saving ducks from certain death?
2. Do you have compassion for others and a concern for animal rights?
3. Are you willing to buy a sterile duck that can be kept as a pet?

If so, consider finding and rescuing a Moulard duck.

Duclair

Duclairs are beautiful ducks. They can be pure white, like a Pekin. They can also be uniform brown or black, with a white speckled chest and a green head. Female Duclairs can have a dark, slate brown head with a creamy back and a brown chest.

Like many other ducks, the Duclair was named after the city where it was bred. Other names include "Canard Duclair" and "Duclair Enten."[201] The Duclair looks similar to the Rouen and the Blue Swedish. Their diet includes "flies, beetles, dragonflies, caddis flies,

[196] Kim Johnson, "What is Foie Gras," *Animal Equality,* Last modified on July 26, 2019, Accessed on April 27, 2021, https://animalequality.org/blog/2019/07/26/what-is-foie-gras/.

[197] https://en.wikipedia.org/wiki/Mulard

[198] https://en.wikipedia.org/wiki/Mulard

[199] https://en.wikipedia.org/wiki/Mulard

[200] "Carrie," *Farm Sanctuary,* Accessed April 27, 2021, https://www.farmsanctuary.org/adopt/adopt-a-farm-animal-carrie/.

[201] "Duclair Duck: Characteristics, Origin, Uses & Full Breed Information," *Roy's Farm,* Last modified on March 11, 2021, Accessed April 27, 2021, https://www.roysfarm.com/duclair-duck/

gastropods, worms, and a wide range of crustaceans... slugs and snails."[202] The breed lays plenty of eggs and has a lot of energy.

Best of all, Duclair ducks are amiable and willing to be tamed.[203] My very first ducks were Duclairs. These beautiful and friendly ducks make ideal backyard pets. If you've never owned ducks before, Duclairs are a good place to start.

Questions to consider:

1. Do you need an excellent, calm pet that can be easily tamed by your children and grandchildren?
2. Are you looking for an excellent forager who can help reduce the insect population of your home and garden?
3. Do you need an excellent source of eggs?

If so, consider buying a Duclair.

Gold Star Hybrid

As you might anticipate from the name, Gold Star Hybrids are beautiful, golden-colored ducks. In addition, the female duck has a dark wing stripe with white stripes on either side. Gold Star Hybrid ducks are excellent for pest control, eating mosquitos that grow in standing water.[204] Gold Star Hybrid ducks are related to Khaki Campbells, but they lay more eggs and have a sweeter personality.[205]

Questions to consider:

1. Are you looking for an excellent forager who can help reduce the insect population of your home and garden?
2. Do you have a mosquito infestation that you need to bring under control?
3. Do you need a sweet pet for your children and grandchildren?
4. Do you need an excellent source of eggs?

If you answered these questions in the affirmative, consider buying Gold Star Hybrid ducks for your backyard collection.

White Star Hybrid

Are you looking for an excellent egg laying duck? Just like the pure-white leghorn chickens who are known for their high-capacity egg laying performance, the White Star Hybrid is

---

[202] https://www.roysfarm.com/duclair-duck/
[203] https://www.roysfarm.com/duclair-duck/
[204] Walter Jeffries, "Gold Star Ducks," *Sugar Mountain Farm,* Last modified on August 12, 2013, Accessed on April 27, 2021, https://sugarmtnfarm.com/2013/08/12/gold-star-ducks/.
[205] https://www.mcmurrayhatchery.com/gold_star_hybrid_duck.html

known for its ability to exceed expectations with egg production. In fact, the White Star Hybrid was expressly bred for this very purpose. Its egg laying capacity is its crowning virtue. White Star Hybrids churn out almost three hundred eggs each year.[206]

Questions to consider:

1. Do you need high volumes of eggs?
2. Do you want to go into business selling hypoallergenic duck eggs from your own backyard?
3. Are you prepared with duck egg cartons and space to store a large volume of eggs?
4. Does your family enjoy eggs?
5. Do you use eggs by the dozen in your own home?

If so, consider buying a White Star Hybrid duck as a constant source of eggs.

Silky Ducks

Have you ever heard of Silkie chickens? These rare, fluffy animals have tufts of soft, fuzz-like feathers. Rather than normal feathers with barbs that hold their feathers together smoothly along the shaft, Silkie chickens have hair-like feathers. This gives them a comical but adorable look. Silkies are tame, small, and fun to play with and cuddle.

But even more rare than Silkie chickens are Silky ducks. Like Silkie chickens, Silky ducks have lacy and hair-like feathers that make them look fluffy. Impress your friends and family with these truly one-of-a-kind pets.

Silky ducks are a rare, beautiful form of bantam ducks. These beautiful animals come in a variety of colors: black, mallard color, white, and dusky. The breed was created by "Darrell Sheraw of Pennsylvania."[207]

Perhaps you have limited space and need a small duck that cannot fly away. Silky ducks do not fly, and they take good care of their eggs.[208] Silky ducks would make a perfect pet for a beginner.[209]

Questions to consider:

[206] "White Star Hybrid," Murray McMurray Hatchery, Accessed April 27, 2021,
https://www.mcmurrayhatchery.com/white_star_hybrid_white_layer_ducks.html.
[207] "Silky Ducks," *Feathersite,* Accessed April 27, 2021,
https://www.feathersite.com/Poultry/Ducks/Silky/BRKSilky.html.
[208] "Silky Ducks," *Feathersite,* Accessed April 27, 2021,
https://www.feathersite.com/Poultry/Ducks/Silky/BRKSilky.html.
[209] "Silkie Duck," *Holderread Farm,* Accessed April 27, 2021,
https://www.holderreadfarm.com/photogallery/silkies_page/silkies_page.htm.

1. Do you need a flightless bird who can graze in the backyard without fear of flying away?
2. Do you need a reliable mother who can incubate duck eggs, whether for business or for pleasure?
3. Do you want to propagate a truly rare type of duck?
4. Are you committed to doing the extra homework to find a breeder who sells this rare type of duck?
5. Are you seeking a small duck that doesn't require as much space or consume as much food?

If so, it's time to purchase your very first Silky duck!

Golden Cascade Duck

The Golden Cascade is a new breed of duck in the United States of America.[210] David Holderread wished for an energetic duck in which it was easy to tell males and females apart early in life. His dream duck also laid profusely and grew quickly.[211] David developed this breed in 1979.[212]

These ducks are a beautiful bronze color with speckles of darker brown. Males have a beautiful, chocolate colored head, a white ring around the neck, and deeper copper on the chest. The females are lighter in color. With two lovely eye stripes and a white neck patch, Golden Cascade ducks can be truly beautiful.

Golden Cascade ducks are on the larger size, weighing up to eight pounds.[213] The ducks lay large, white eggs that can weigh 75 grams.[214]

Questions to consider:

1. Do you need an excellent source of large meat birds?
2. Is it important to you that you know which gender of duck you are purchasing?
3. Do you need a good source of large eggs?
4. Are you looking for a beautiful gold-colored duck?

---

[210] "Golden Cascade Duck," Last modified on July 3, 2016, Accessed on April 27, 2021, https://www.breedslist.com/golden-cascade-duck.htm.
[211] "Golden Cascade Duck," Last modified on July 3, 2016, Accessed on April 27, 2021, https://www.breedslist.com/golden-cascade-duck.htm.
[212] "Golden Cascade Duck," Last modified on July 3, 2016, Accessed on April 27, 2021, https://www.breedslist.com/golden-cascade-duck.htm.
[213] "Golden Cascade," *Wikipedia,* Accessed April 27, 2021, https://en.wikipedia.org/wiki/Golden_Cascade.
[214] "Golden Cascade Duck: Characteristics, Uses & Full Breed Information," *Roy's Farm,* Last modified on March 11, 2021, Accessed April 27, 2021, https://www.roysfarm.com/golden-cascade-duck/.

If so, consider buying the Golden Cascade Duck for your backyard farm.

Chapter 4: Bringing Your Duck Home: Finding the Right Supplier

Now that you have all the information you need about duck breeds, it is time to make your final choice.

Depending on which breed you decide to buy, you will need to consider availability, suppliers, and breeders. Some breeds are readily available throughout the United States, while others will be more difficult to find.

For example, perhaps you chose to buy a Pekin duckling. If so, you are in luck. This common duck breed is readily available at any local Tractor Supply Company outlet.

But maybe you were entranced by the beauty and grace of a rarer breed, such as the Miniature Silver Appleyard ducks. According to some sources, this breed is only sold by two suppliers in the United States. Wheeler Farms supplies this rare breed.

Some people choose to source their ducklings from individuals and private breeders in their area. Others choose to order from a reputable online breeder. Still others network with friends and family until they find the animals they are looking for.

Let's look at a couple options you have in finding your perfect duck.

Social Networking to Find Private Breeders

As you search for an excellent source for the breed you have selected, visit Facebook pages and find other breed enthusiasts like yourself. Each breed has its own page. On this page, you will meet others who are buying, selling, and breeding the breed you have selected. You can ask questions and receive answers about reputable breeders in your area.

Here are a few Facebook pages to get you started.

1. Indian Runner Duck Facebook Page: https://www.facebook.com/groups/indianrunnerducksusa/
2. Another Runner Duck Facebook Page: https://www.facebook.com/groups/2226765199
3. Miniature Silver Appleyard Facebook Page: https://www.facebook.com/groups/364416147091676
4. Full Size Silver Appleyard Duck Facebook Group: https://www.facebook.com/groups/251643688377925

5. Rouen Duck Facebook page: https://www.facebook.com/groups/125384880859303

Individuals turn to social media to advertise their small-scale breeding operations. They may advertise within the duck breed Facebook page. In addition, they may have their own Facebook page that represents their small business. They may also use local Facebook poultry groups to advertise their breeding operation.

Search your social media for local poultry groups where small-scale breeders may be advertising their hatching eggs, ducklings, and full-size ducks.

For example, here is a page where Kansas poultry owners may buy, sell, and trade their ducks, chicken, and poultry: https://www.facebook.com/groups/354174588004868/ Your area may have a similar page that puts you in contact with other owners.

When considering buying from a small-scale breeder, consider the following questions:

1. Is it important to me to support small, local businesses?
2. Am I comfortable meeting up with people I do not know to obtain eggs or ducklings?
3. Have I done my research?
4. Do I understand the background and quality that this breeder represents?
5. Is it important to me to obtain live birds from the area that will not need to be shipped?
6. Do I have the information I need to determine whether the birds are healthy?

If you answer "yes" to these questions, consider shopping on social media for an excellent local breeder in your area. A poultry swap is another place to find private breeders.

Attending Poultry Swap Groups and Fairs

A Poultry Swap is a huge outdoor meeting where many different breeds of chickens, ducks, and other poultry are up for sale. There are often cages, butchering equipment, nesting boxes, and other poultry equipment for sale in the swap group.[215] Think of a swap group as a Farmer's Market for poultry and poultry supplies. You can browse, look around at various booths, and decide whether or not you want to buy anything.

---

[215] "What to expect at a chicken swap (you might be surprised!)" *Murano Chicken,* Accessed April 27, 2021, https://www.muranochickenfarm.com/2014/05/what-to-expect-at-chicken-swap.html.

How do you know the quality of the animals you can buy in a Poultry Swap Group? The quality of poultry you will find at a swap group can vary widely. Some vendors sell show quality animals, while others are just backyard farmers who want to get rid of some extra animals.[216] Like always when buying an item, check the background and credentials of the buyer. Take a careful look at the duck you are planning to buy.

How do you find a swap group in your area? Many states have their own Poultry Swap pages on social media. By joining these Facebook pages, you will receive information on live events in your vicinity that you might be interested in attending.

Questions to consider:

1. Do I enjoy meeting other poultry breeders?
2. Do I live near a large city that would have a major poultry meet-up?
3. Do I want to exchange information with other people who have a passion similar to my own?
4. Do I have hatching eggs, ducklings, equipment, cages, or adult ducks that I want to sell?

If so, a Poultry Swap would be an excellent place to find new ducks for your backyard flock.

Talking to Friends and Family to Find Trustworthy, Local Breeders

When looking for a trustworthy breeder, word of mouth is an important thing to consider. Do you have friends who raise ducks or poultry? They might have personal connections with trustworthy breeders. Perhaps your friends themselves would be willing to give you or sell you some hatching eggs or ducklings.

Ask around on your personal Facebook feed. Find personal friends who are knowledgeable about backyard animals. Use the grapevine to get good information about who to trust.

Questions to consider:

1. Is it important to you to find a breeder you can trust who comes with good recommendations?
2. Do you want to make connections with local, family-based farmers who are personally known to your friends?
3. Do you want to establish relationships with others who are backyard farmers like yourself?

---

216 "What to expect at a chicken swap (you might be surprised!)"

If so, reaching out to your personal friends and acquaintances can be an excellent place to start when searching for new ducklings for your flock.

Buying from a Local Feed Store

Another option is to buy your ducklings from a local feed store. Tractor Supply Company is well known for selling hundreds of chicks and ducklings each year. During the pandemic, feed stores saw dramatic increases in poultry purchases.[217]

Some local feed stores are excellent sources of trustworthy information. They can answer your questions and help you find what you need to get started raising backyard ducks.[218] Personally, I have used the same feed store for over twenty years. The owners know me as a person, and I know them. When I had a question about poultry feed, they answered my question honestly and knowledgeably—even when it meant that I realized that I did not need to buy another bag of feed that day.

A downside to buying chicks from a local feed store is that they are often under socialized. In other words, they have been cooed over by dozens or hundreds of people. They have been kept in holding tanks with hundreds of other chicks. They have been stressed by Tractor Supply workers quickly and carelessly grabbing chicks for potential customers. They have not been personally handled and are not as easy to tame.

Questions to consider:

1. Do you want a quick and easy option for buying new ducklings?
2. Do you want to interact in person with knowledgeable poultry experts?
3. Do you want to inspect the chicks yourself, rather than getting the run of the lot from the online breeder?
4. Are you raising animals for eggs or meat, and it is not important to you to have a duck that has been handled extensively?

If so, using a local feed store can be an excellent option for purchasing your next batch of ducklings.

---

[217] Bill Chappell, "'We Are Swamped': Coronavirus Propels Interest In Raising Backyard Chickens For Eggs," Last modified on April 3, 2020, Accessed on May 6, 2021, https://www.npr.org/2020/04/03/826925180/we-are-swamped-coronavirus-propels-interest-in-raising-backyard-chickens-for-egg.
[218] Jason Roberts, "17 Best Hatcheries to Buy Chickens Online," Last modified on April 25, 2020, Accessed on May 6, 2021, https://www.knowyourchickens.com/buy-chickens-online.

Using a Mail Order Breeder

Sometimes, it is nearly impossible to find the breed you are looking for in a particular locale. Rare breeds, such as the Silky duck or even the Silver Appleyard Bantam, may be difficult to find in your area. If this is the case, you may want to research mail order hatcheries. These hatcheries will mail your hatching eggs, young ducklings, or even adult animals through the United States Postal Service.

Just like any other duck buying option, buying from an online breeder has its pros and cons.

Pros:

1. With an online breeder, you often get a better price. Sometimes, buying from an online company can give you a better price.[219] But other times, local breeders who are just getting started with their duck breeding business are willing to give you a more competitive price. Do your research.
2. With an online breeder, it is often easier to find rare breeds. As mentioned above, local breeders do not always branch out into rare breeds such as Silky ducks. Online hatcheries, however, are able to specialize.
3. With an online breeder, you can often get insurance for the gender of your ducklings.[220] If you have been promised female ducks and you get male ducks instead, the breeder may be willing to refund part of your money. Small local breeders often do not offer insurance. If the ducklings you purchase are not the gender you had hoped for, too bad so sad. However, with an online breeder, you can often purchase insurance for your animals.

Cons:

4. With an online breeder, there is often a higher mortality rate due to stresses during shipping. Ducklings shipped in the mail are more likely to die soon after arrival.[221]
5. With an online breeder, the young birds have to endure the stress of shipping. Not only does this stress affect their health, but it also affects their personality and well-being. Stressed birds are less likely to bond with their new owners.

---

[219] Jason Roberts, "17 Best Hatcheries to Buy Chickens Online," Last modified on April 25, 2020, Accessed on May 6, 2021, https://www.knowyourchickens.com/buy-chickens-online.
[220] Jason Roberts, "17 Best Hatcheries to Buy Chickens Online," Last modified on April 25, 2020, Accessed on May 6, 2021, https://www.knowyourchickens.com/buy-chickens-online.
[221] Jason Roberts, "17 Best Hatcheries to Buy Chickens Online," Last modified on April 25, 2020, Accessed on May 6, 2021, https://www.knowyourchickens.com/buy-chickens-online.

6. With an online breeder, you have no personal contact with the breeder. This means you have less information about the duckling's history, parents, breeding information, and habits. You don't have a personal touch point that can inform you about the conditions the bird was raised in and what type of environment it is accustomed to.

7. With an online breeder, the duckling is less likely to be tame. The duckling that is shipped through the mail has likely not been personally handled by a loving breeder, and it is more difficult to tame the birds and win their trust.

8. With an online breeder, you can't put your mind at rest about whether or not the birds were raised in ethical surroundings. Unless you choose a small, family-owned online hatchery, such as J M Hatchery (see below), you have no guarantee that your ducklings were raised ethically with minimal mistreatment.[222] You have no guarantees that the hatchery has avoided unnecessary animal death when raising the poultry.[223]

9. With an online breeder, you usually have to buy a minimum.[224] Sometimes, you may have to order 25 ducklings at a time; other times, you have to order at least six birds. With a local breeder, the minimum is usually two. If you are starting a very small operation, you may want to opt for a local breeder.

Here are a few of the most common breeders in the United States, recommended by Jason Roberts.[225]

1. Mt. Healthy Hatcheries. Mt. Healthy Hatcheries boasts that they are the "home of the healthiest chicks." They also offer "the cutest bills you'll ever get!" Get it? Their offerings include Mallards, Khaki Campbells, and White Pekins.

2. Holderread Farm. Holderread Farm ships adult birds as well as ducklings. They ship a variety of more rare versions that are difficult to find elsewhere. For example, they sell Large Silver Appleyard ducks, Miniature Chestnut Appleyard ducks, Call ducks (butterscotch, gray, white, and snowy), various colors of the rare Dutch Hookbill (dusky, white bibbed dusky, and white),

---

[222] Jason Roberts, "17 Best Hatcheries to Buy Chickens Online," Last modified on April 25, 2020, Accessed on May 6, 2021, https://www.knowyourchickens.com/buy-chickens-online.

[223] Jason Roberts, "17 Best Hatcheries to Buy Chickens Online," Last modified on April 25, 2020, Accessed on May 6, 2021, https://www.knowyourchickens.com/buy-chickens-online.

[224] Ibid

[225] Jason Roberts, "17 Best Hatcheries to Buy Chickens Online," Last modified on April 25, 2020, Accessed on May 6, 2021, https://www.knowyourchickens.com/buy-chickens-online.

Saxony, and Welsh Harlequin. They also carry a rare kind of duck called the Silky duck.

3. JM Hatchery. JM Hatchery is a faith-based hatchery in Pennsylvania. The hatchery offers White Muscovy ducks and Khaki Campbell ducklings. The business has been in the family for generations. Their mission statement says, "We strive to honor our customers with a quality product, excellent customer service, and a wonderful experience dealing with JM Hatchery. It is our wish that each of our customers would prosper, not only materially, but also spiritually by coming to know our Savior in a personal living relationship."[226] Buying from this hatchery, you will draw on a wealth of personal, family, and generational investment in your future flock.

4. Meyer Hatchery. Meyer Hatchery offers a wide selection of birds you can choose from. Their offerings include Buff, Blue Swedish, Cayuga, Black Swedish, Mallard, White Layer, White Muscovy, Jumbo Pekin, Rouen, White Pekin, Black Runner, Chocolate Runner, Blue Runner, Saxony, Fawn Runner, White Runner, Welsh Harlequin, Silver Appleyard, White Crested, Grimaud Hybrid, and Duclair ducks.

5. Murray McMurray Hatchery. Murray McMurray Hatchery sells all types of ducks, including White Crested, Welsh Harlequin, White Star Hybrid, Silver Appleyard Full Size, Saxony, Runner, Pekin, Rouen, Muscovy, Jumbo Pekin, Khaki Campbell Gold Star Hybrid, Mallard, Duclair, Cayuga, Buff, Blue Swedish, and Black Swedish.

6. Cackle Hatchery. Cackle Hatchery offers White Pekin, Rouen, Khaki Campbell, Welsh Harlequin, Ancona, Black Swedish, Cayuga, Fawn and White Runner, Blue Runner, Black Runner, Buff, Golden 300 Hybrid Layer, Jumbo Pekin, Mallard, White Crested, White Layer, and Call ducks.

7. Stromberg's. Stromberg's sells a wide variety of ducklings, including assorted Runner, Black Runner, Black Swedish, Blue Runner, Blue Swedish, Buff, Cayuga, Chocolate Runner, Fawn Runner, Khaki Campbell, Magpie, Mallard, Crested White, Rouen, and White Pekin. They also offer adult Mandarin ducks for over $200 a pair.

8. Freedom Ranger Hatchery. Freedom Ranger Hatchery offers Khaki Campbell and White Muscovy ducks.

9. Hoover's Hatchery. Hoover's Hatchery offers a free duckling care guide to help you get started on your duck-raising adventure. It also sells Pekin, Mallard, Khaki Campbell, Indian (Fawn and White) Runner, Rouen, Buff, Blue Swedish, Cayuga, Blue Runner, Golden 300 Hybrids, Welsh Harlequin, Black Swedish, Saxony, Ancona, Black Runner, Chocolate

---

[226] "About Us and Our Faith," JM Hatchery, Accessed April 27, 2021, https://jmhatchery.com/about-us-and-our-faith/.

Runner, Magpie, Muscovy, Silver Appleyard, White Crested, White Layer, and Golden Cascade ducks.

10. Privett Hatchery, Inc. Privett Hatchery offers Black and Blue Runner, Domesticated Grey Mallard, Chocolate Runner, Khaki Campbell, Magpie, and White Pekin ducks.
11. Ideal Poultry. Ideal Poultry offers quite a few kinds of domestic waterfowl. Buff, Crested Buff, Cayuga, Crested Cayuga, Khaki Campbell, Crested Khaki Campbell, Black and White Magpie, Crested Black and White Magpie, Domesticated Grey Mallard, Crested Domesticated Gray Mallard, Snowy Mallard, White Pekin, Crested White Pekin, Rouen, Crested Rouen, Fawn and White Runner, Crested Fawn and White Runner, Blue Swedish, and Crested Blue Swedish ducks. What a variety of crested and un-crested ducks!
12. Town Line Poultry Farm. Town Line Poultry Farm offers Cayuga, Indian Runner, Khaki Campbell, Rouen, and White Pekin ducks.
13. Ridgway Hatchery. Ridgway Hatchery has bred their own variety of White Pekins, which they call Ridgway Mammoth White Pekins. They also offer Indian Runners, Ridgway Rouens, Blue Swedish, and Ridgway Genuine Flying Mallards.
14. Chickens for Backyards. Not limiting their supply to chickens, Chickens for Backyards supplies a few varieties of ducklings, including Cayuga, Rouen, White Pekin, Ancona, Khaki Campbell, Blue Swedish, Welsh Harlequin, Black Swedish, and Fawn and White Runners.
15. My Pet Chicken. My Pet Chicken, ironically, also sells ducks. Their offerings include Black Runner, Black Swedish, Blue Runner, Blue Swedish, Buff, Chocolate Runner, Cayuga, Fawn and White Runners, Golden 300 Hybrid Layer ducks, Grimaud Hybrid Pekin, Jumbo Pekin, Khaki Campbell, Mallard, Muscovy, and Pekin ducks.
16. Welp Hatchery. Welp Hatchery offers Crested, Cayuga, Buff, Dark Rouen, Domesticated Gray Mallard, Swedish, Runner ducks of all colors, Giant White Pekins, Khaki Campbells, and Magpie ducks.
17. Purely Poultry. Purely Poultry offers Mallards, Call ducks, Grimaud Hybrid Pekins, Pekins, Cayugas, and Khaki Campbells.

When you choose a mail order hatchery, it's critical to do some research. Don't just jump on the one with the best price. Instead, read reviews. Talk to friends and family. Find out all you can about the quality of the hatchery. One friend I know bought her poultry from a hatchery that offered very good deals, low prices, and very low shipping costs. When I heard about the deal my friend found, it seemed too good to be true. However, she assured me that it was for real. A few days later, she received the healthy baby poultry from the hatchery. It seemed that the online hatchery was reliable after all.

However, in a few months, she realized that the growing birds were not high quality, well-bred animals. They were scraggly, scrawny, and just plain ugly. In other words, the birds themselves were alive and healthy, but they were not well bred or high-quality poultry in the first place.

Be careful when ordering from online breeders. Don't just read the descriptions on the site; check out references and recommendations from real life friends and poultry breeders.

Questions to consider:

1. Am I starting a large operation with many birds?
2. Am I looking for value over tameness?
3. Am I okay with losing a few birds during shipping?
4. Is it important to me to know the gender of the ducks?
5. Do I want efficient service that does not require travelling to meet up with a stranger?
6. Have I done my research about the quality of the online breeder?
7. Am I looking for a rare variety of duck that is not readily available in my area?

If so, you may want to consider purchasing your hatching eggs or ducklings from an online hatchery.

How many ducklings should I buy?

To answer that question, you need to ask yourself what your purpose is in raising ducks. Also take into consideration how much space you have, how close you live to neighbors, how tolerant your neighbors are to noise and smell, and how prepared you are for a duck start-up.

Here are some factors to consider when purchasing ducks:

1. If you are a new duck owner, you might want to start with two ducks. Don't buy only one; it will become lonely and depressed. However, two ducks can be quite happy together. If you buy just one pair of ducks, you can get your feet wet. You can gain experience with keeping them healthy and happy. You can experiment with the swimming pool, water setup, feed bill, space requirements, and clean-up that is necessary. If you discover that you enjoy ducks and are well acquainted with how much work they are, you can expand your flock.

2. If you have a lot of space, you can buy a bigger flock. If you live on the outskirts of town, don't have nearby neighbors, and have a large backyard pond with a cleaning element, you don't have to worry so much about your ducks. They will take care of themselves, bathing in the pond.
3. If you have quite a few natural predators in your area, you might want to try buying a few extras. Inevitably, you will lose some ducks to raccoons, foxes, possums, or stray dogs who are roaming the area. If you want to end up with a good number of ducks after predation has occurred, be generous when buying them.
4. If you have intentions of starting a commercial duck breeding operation in your backyard, you might want to consider buying more ducks to begin with. If you plan to sell duck meat or eggs, you can consider buying more to begin with as well.
5. If you want to buy from an online breeder but don't want to purchase a large flock, consider making a joint order with other friends or family. In other words, several birds can be spoken for by each friend. After you receive the order, you pass out the ducklings and are reimbursed for the money you spent on the order.

Common Mistakes New Duck Owners Make When Purchasing—and How to Avoid Them.

As a first-time duck owner, making mistakes comes with the territory. You will learn from your mistakes and be all the better for it. However, the more prepared you can be the better.

Here are a few common mistakes first-time duck owners often make.

Making a decision at Tractor Supply. Perhaps you know the feeling. You went to Tractor Supply Company one spring morning, just as you were emerging from a long, cold winter. You needed to pick up some dog food or you'd heard of a great sale on warm, winter coats. It's been a long winter, and you've been frustrated with your kids and spouse and life in general.

Suddenly, you hear a soft peeping sound coming from the back of the store. Hurrying over, you see dozens of fuzzy yellow ducklings milling around. Your heart melts. You just have to take a few home to your children. Just imagine the look on your kids' faces when you show them their brand-new pets!

Energized by a new passion you didn't know you had, you rush around the store, buying the supplies you know you'll need. A brooder box, some duckling feed, a poultry feeder and waterer. Soon, you have your very own, brand-new ducklings secured in an adorable cardboard box.

For several weeks, you post about your brand-new pets all over social media. Your kids love playing with them in the yard and watching them swim. You know you're not just one of "those people" who don't even secure the coop or provide proper accommodations. So you invest in a ready-made coop, a large wading pool, and several other pieces of equipment.

But when the newness wears off, you wonder what in the world you just did.

Investing too much in the ducks before you know you're serious about them is a very common mistake that prospective duck farmers often make. Duck owners joke about going to Tractor Supply during chick season. But in all seriousness, it's best to never buy on an impulse—especially not ducks. If you're not one hundred percent certain that you can handle the responsibilities of duck ownership, don't spend a lot of money on equipment. And certainly, don't purchase the ducks themselves. They'll run the risk of becoming another statistic eaten by predators on a city park pond. That's why it's important to read books like this thoroughly before you make your decision. Make a commitment to yourself to never buy a duck "just because" you saw them in Tractor Supply.

As we discovered above, Tractor Supply can be an amazing option for buying ducklings. As part of a wise, measured decision, buying at Tractor Supply can be a wonderful choice. With your spouse and other household members, decide how many birds to buy, and then stick to that number, even if you're tempted to bring home more once you get to the store. Make a commitment now to never buy on impulse just because they looked cute.

Not being prepared to invest financially. If you're going to have ducks, you're going to spend money. Period. You might as well get used to the fact that these animals are going to cost you something. You'll have to continually buy bedding, straw, pine chips, food, hoses, water containers, de-icing buckets, and other supplies. If you're especially dedicated (or have more money than time and energy), you may find yourself buying a complicated self-filtering water pond, a giant stock tank, or going the extra mile to lay down different types of stone, rock, sand, or other drainage facilties.

Don't buy ducks on a budget or try the do-it-yourself version. Some of these hacks can work for a while, but in the long run, you're going to end up buying the expensive version. You might as well just do it right now. Yes, do your research and make sure you're buying high-quality equipment that will stand the test of time. But you can just give up on being thrifty.

Not being prepared to manage the water and mud. Water and mud are two of the biggest enemies you'll fight with over the next few months, so you need to be prepared.

We'll talk about bedding, water, and mud below. But as you consider your duck purchase, it's important to be ready as early as possible. The better prepared you are, the less you'll suffer later on down the road.

Chapter 5: Preparing the Coop

Now that you have made a decision about the type, breed, and number of ducks you want to buy, it's time to prepare the coop. As discussed above, the coop style you choose depends on the unique characteristics of the duck in question. Flighty ducks who like to escape the confines of their pens need a tall coop with a roof so they cannot fly out. If you decide to buy Muscovy ducks, consider the fact that they will need a roost. Large ducks need more space, while smaller ducks can be kept in a smaller area.

Review the above information about the varying needs of different duck breeds. Then, create a custom designed coop that will be perfect for the new residents of your backyard.

Basic elements of a coop. The basic elements of a coop include an outdoor run and a shelter. The outdoor run can be hand built from fencing. You can purchase the entire set up from Tractor Supply Company or another feed store. You could order a pre-built poultry coop from Amazon.

You can choose to connect an outdoor coop to an indoor shelter through a hole or door in the coop. Or you can place the shelter inside the coop itself to provide less openings through which predators could potentially enter.

For the outer run, you can consider starting with a basic dog run and adding features. The shelter can be handmade, or you can be more creative. Could you use a doghouse igloo? A prefabricated shed? Part of the barn? Ducks are happy, contented creatures and are usually not too demanding about their space. We'll look below at a few more details and requirements for the indoor shelter.

How much space per duck? Sources agree that each duck needs four square feet.[227] A duck pen should be at least 24 square feet in total, even if you have fewer than six ducks. Ducks need space to roam around, explore, and enjoy their surroundings.

Building a predator-safe coop. Ducks have many natural predators, and they are mostly defenseless. Most chickens can fly away from predators, but many ducks do not have this natural skill. Instead, most ducks are confined to waddling awkwardly and helplessly on the ground, quacking comically as they run from predators. Predators of the duck include raccoons, coyotes, badgers, foxes, skunks, mink, corvids, hawks, gulls, owls, eagles, and even squirrels.[228] The latter animals are known mostly for eating duck eggs.

---

[227] Lisa Steele, "A Guide to Duck Houses," *HGTV,* Accessed on May 7, 2021, https://www.hgtv.com/outdoors/gardens/animals-and-wildlife/a-guide-to-duck-houses.
[228] "Top Duck-Craving Predators," *Delta Waterfowl,* Accessed May 7, 2021, https://deltawaterfowl.org/top-duck-craving-predators/.

Predators can enter coops in the most unexpected and ingenious ways. Our first flock of poultry was decimated when dogs entered our fenced yard, made their way into our further fenced poultry yard, and killed every bird. To fully protect against predators, you need the following elements:

1. Chicken wire or fencing with small enough holes. Predators are sneaky, and they can reach their paws into fences and grab hapless birds who are walking by. Additionally, wild birds can enter coops and bring diseases. The finer the mesh, the more critters you can prevent from bothering your ducks.

2. Roof. Not only can predators use their hands to reach and grab birds, but many types of predators can also climb. Raccoons and possums can scale fences and enter the cage from above to feast on helpless ducks. Hawks can fly in from above and attack birds from the unprotected sky zone. The roof must be as secure as you know how. A roof made of chicken wire or fencing, securely attached on all sides to the coop walls, should be secure enough to keep out invading predators.

3. Underground protection. You may feel like you have covered all the bases. You have used fine mesh chicken wire and covered the top. However, predators can also dig. You must ensure that dogs and other diggers cannot enter your coop from below. Try creating your outdoor poultry yard out of a dog run or other fencing. Then, after creating a secure roof, dig underground at least a foot and attach an additional segment of wire under the ground. Ensure that the additional wiring is securely attached to the main coop fencing.

4. Consider joints. By joints, I mean the places where one part of your coop set up attaches to another. The connection between a coop and a house can be a key place where predators can enter. Never underestimate the ability of predators to enter the most seemingly secure house. After our first flock of poultry was eaten by dogs, we built a coop that has stood the test of time. For over twenty years, this coop has kept out all types of animals. One of the most important aspects of this secure coop is the secure joints. Where the coop opens into the house, we attached layers of chicken wire that were secured inside the house and wrapped around the joint to be secured onto the inside of the coop as well. In other words, there were no exposed areas where racoons and other predators could sneak in.

Drainage. When you are building a coop for your ducks, you must remember a factor that is commonly overlooked by new duck owners. One of the most important things to keep in mind when constructing a duck living space is drainage. If you've had experience raising chickens or other kinds of poultry, you likely did not have to take drainage into consideration. But with ducks, it will become an absolute necessity.

While raising ducks, you will be going through a lot of water in your coop. Every morning, you will be dumping out dirty water, hosing out the ducks' water bowl, and refilling it. If you choose to have a baby pool in your coop, you will need to have a place to dump that as well. After a few weeks or months of duck ownership, you will realize that it is critical to allow the water to drain away from the duck pen rather than into it. Ducks thrive on water and mud, and they would absolutely love to live in pig pen conditions. However, most duck owners have other opinions. We know that too much mud can actually damage ducks' feathers and cause a disease called Wet Feather. In addition, mud is inconvenient for humans who are trying to care for the ducks. Few people enjoy slopping through knee-deep mud every day on their way to feeding and watering the ducks.

Your coop needs to have a spacious gate or door through which you can comfortably enter to deal with the ducks' daily water needs. The duck coop needs to have a large enough opening to accommodate the pool, water bowls, hose, and other necessities of duck ownership.

You will need to think about how to bring water safely to the coop, as well as how to siphon water economically and efficiently away from the coop. You want to eliminate as much mud as possible. The rainwater or splashed drinking water that enters the coop should drain away from the ducks' feet and feathers as fast as possible, and never come back. Basically, you want to avoid creating a pig pen.

Some duck owners have designed a drainage pit, a drainage pipe, or a drainage ditch that leads away from the coop.

1. Drainage pit inside the coop. You can dig a very deep hole in the coop and reinforce it with chicken wire on the inside. Place secure mesh over the top of the pit. Drain your daily water waste through the mesh covering into the drainage pit every day. Make sure to secure the covering so that ducks do not fall into the hole. The drainage pit can be open, or you can fill the pit with well-draining material such as gravel or sand.
2. Drainage trench that leads away from the coop. Similarly, you can create a trench that leads downhill, continues across the coop, goes through the coop's underground fence, and leads into a drainage area far away from the duck coop. Ensure that the place where the trench exits the coop is secure against predators. Cover the trench with wire mesh so that the ducks are not constantly digging and swimming in the trench. Another option for the drainage trench is to fill it with gravel and sand. Ensure that the top is still covered so that the ducks do not hurt their feet on the gravel.

3. Drainage pit outside the coop. In the limited space in my backyard, I have found the easiest solution is to dig a large hole just outside the wire mesh of their coop. Every morning, I pour the water through the fence into the hole. I manage erosion by creating channels that lead toward the exit. In this way, the water goes the right direction and does not flood the rest of the coop.
4. Drainpipes. You can set up a complete duck drainage system, complete with drainpipes that lead outside the coop to a pond or a garden.
5. Building on a slope. A natural hill or slope in your backyard can be used to great advantage when building your chicken/duck coop. If the ground is naturally sloped, then you will be able to easily and naturally drain water away from your ducks.[229] A slope is also an excellent asset that you can use when building a duck pond. See below for more information on building your duck pond.

When working on drainage, here are some things to keep in mind:

1. Keep in mind that chickens and ducks frequently change the surface of their coop, impacting the direction water drains away from your coop. If your coop is covered in sand, dirt, straw, or another type of substrate, the birds themselves will change the water flow and erosion patterns of the dirt. Ducks will dig holes with their beaks, and chickens scratch away straw and dirt. Chickens will also create pits for themselves to dust bathe in. These elements will affect the water flow of your coop. In other words, you may have dug surface channels to direct the water where you want it to go. But one day later, your chickens and ducks may have interfered with those channels so that the water no longer flows in the direction you want it to.
2. Keep in mind that drainpipes can easily become clogged with mud, manure, straw, and sand. Ensure that you have a large, predator-secure drainpipe that will allow straw to pass through it.
3. Any pit, trench, fencing, or drainpipe will need to be cleaned regularly. Ditches and pits will start to fill up with mud and organic matter, because the water you pour into them will be full of mud and organic matter. Every now and then, remove the cover from the drainage pit and clean out all the debris. Drainpipes must also be cleaned. Remove the mud, manure, straw, and sand that clog them.

Preparing the substrate. When I first started raising ducks, I raised them in the natural dirt coop that I had kept my chickens in for many years. However, I soon realized

---

[229] "Self-Draining Duck Pen—Is It Possible?" *Backyard Chickens,* Lost modified on April 11, 2014, Accessed on May 7, 2021, https://www.backyardchickens.com/threads/self-draining-duck-pen-is-it-possible.872928/.

that this was a mistake. Endless mud deteriorated their feathers and caused a hazard for their health.

I have found practical solutions to this problem in several ways.

1. Laying landscaping cloth. First, I dug deep and lay down landscaping cloth. The purpose of the landscaping cloth was to keep the sand and gravel from mixing with the dirt in the coop. I ensured that the landscaping cloth was laid down in such a way that it drained toward the drainage pit I had created near the edge of the coop. It was important to me that when water drained down below the surface, it would be directed toward the drainage pit rather than toward the edges of the coop or the middle of the coop.
2. Laying down gravel and sand. On top of the landscaping cloth, I lay down gravel and sand. Typically, course gravel should be laid down at the bottom, followed by finer gravel and sand. Courser, sharper gravel can damage the delicate webbing between ducks' feet, so it's important to keep the cage free from sharp and damaging material.

This type of substrate was effective to reduce mud for a while. Eventually, the straw, leaves, and other organic matter I added on top of the sand mixed with the sand and became a rich, loose organic compost substance. However, it is still looser and easier to drain than the compact dirt that was in my chicken coop at first.

Unlike chicken poop, duck poop cannot be raked out of sand. Instead, duck poop is slimy and mixes with the sand. Your biggest danger is that ducks will sit in their own poop, become slimy, get damaged feathers, and get Wet Feather. I used straw and leaves on top of the sand to avoid this problem. Every few days, I added additional straw and leaves to keep the surface of the pen clean and dry. This method works well on top of sand. However, if you add straw and leaves on top of mud, it will generally mix with the mud and create a very strong, viscous concrete substance. If you keep chickens with your ducks, the chickens can scratch in the sand, leaves, and straw to mix it up for you. This will help you avoid the constant presence of poop on the surface of the ground. I always keep chickens with my ducks. The few weeks that I kept my ducks separate from my hens, I instantly realized my mistake. The rate at which poop accumulated on top of my fresh straw and bedding was unbelievable. Chickens have been a lifesaver for my ducks' coop cleanliness. In addition, the sand did provide better drainage and less mud than the dirt itself had.

In addition to sand or gravel, you can also choose to lay down concrete at the base of your outdoor pen. Concrete effectively keeps out digging predators. In addition, a concrete base can be easily hosed off! This is a wonderful plus if you are tired of constantly digging, scraping, mixing, and scooping poop. Try building your concrete slab at a slight angle so your rinse water easily drains away. You can place straw or other bedding on top of the concrete.[230]

Think creatively. You can line the floor of your coop with rubber stall mats: https://www.rubberflooringinc.com/weight-room-mats.html

These mats are also easy to rinse and can have gaps in between them where the water can drain into the ground below. It's important to rinse them frequently, though. If you don't keep them clean, poop can accumulate on the surface just as easily as it can accumulate on dirt, straw, or sand. Make sure to rinse the stall mats daily and cover them with straw or another kind of bedding to keep your ducks happy.

Building a roof on the outdoor run. As mentioned above, it's important to have a roof over your coop to keep out predators. If you want your coop to be open to the elements (sun, rain, and snow), you can cover the top of your outdoor coop with chicken wire or fencing.

However, if you would like to keep your coop as dry as possible, there are a variety of materials that you can use to create a roof over your coop.

1. A tarp can be a low-cost option for keeping rain out of the coop. You can also add panels of corrugated polycarbonate.
2. Clear corrugated polycarbonate allows the sun to enter the coop, but keeps out rain, snow, and excessive moisture.
3. In addition, you can cover the top of your outdoor coop with sheet metal.
4. Many stores sell kennel roofs for use with a dog run. If using a kennel roof, keep in mind that ducks and chickens have additional predator protection needs that were not taken into consideration when building a kennel roof for a dog. Please ensure that you cover the triangle under the roof with appropriate netting or wire mesh.

Personally, I first covered the top of my coop with wire mesh, and then I added a wooden frame that held sloping sheets of clear corrugated polycarbonate. In this way, we keep our animal coop free of rain but allow sunshine to enter freely so the coop does not get too dark.

Providing Appropriate Shelter for Your Ducks

---

[230] "Duck coop on concrete slab??" *Backyard Chickens,* Last modified on May 31, 2011, Accessed on May 7, 2021, https://www.backyardchickens.com/threads/duck-coop-on-concrete-slab.515211/

After you have prepared the duck coop, it's time to turn your attention to a safe, warm, and secure indoor shelter for the duck. Many duck experts agree that ducks do not need as much warmth and shelter as chickens. The Cape Coop explains, "They don't need much in terms of shelter—just a secure, safe place to retreat to. It can be a sectioned off corner of your barn or you can even house them in your chicken coop if you want to."[231] Some sources say that a three-sided area with a roof is sufficient for ducks. However, it is important to understand that ducks need to be sheltered from the wind and weather.

When building a duck house, consider what small shelters you may already have on hand. Do you have a barn where they can sleep? A shed where they can go inside at night? How about a doghouse or an old box? Each of these can make an excellent shelter for a duck.[232]

When considering making a duck house, it's important to keep in mind that ducks create a lot of humidity. Provide them with ample ventilation[233] so that the house doesn't become stuffy, or moldy.

It's also important to remember that ducks are awkward and unwieldy.[234] Make sure the door to your duck house is not too small. Ducks will waste a lot of time trying to barge their way through small chicken house doors. You don't want your duck to hurt himself or herself when trying to wiggle through a too-small opening. An ideal size for the door is 14 inches by 14 inches.

Ensure that your duck house doesn't have any nails or wires sticking out that might hurt a duck's tender webbed feet. If you followed my instructions about securing the joint between the house and the yard, you may end up with occasional loose wire ends or pieces of nail. These need to be completely covered and sealed so that they don't tear the ducks' feet. Torn, lacerated, or punctured feet can lead to infections such as bumblefoot.

You may choose to make a ramp that leads up into the duck house. If you keep your ducks with your chickens, you may already have a ramp into your coop. Ramps are excellent for ducks but ensure that wooden ramps do not have any protruding nails or splintery wood that will harm ducks' feet.

Furthermore, make sure that the ramp has traction on it so that ducks don't have trouble with slipping off. With their soft, wet, slippery feet, ducks do not have as much traction as chickens do.[235]

---

[231] "How to Make a Duck House," *The Cape Coop,* Accessed May 19, 2021, https://thecapecoop.com/make-duck-house/
[232] "How to Make a Duck House"
[233] "How to Make a Duck House"
[234] "How to Make a Duck House"
[235] "How to Make a Duck House"

The Cape Coop provides an amazing, step by step explanation of how Liz built her first duck house. Follow their instructions here: https://thecapecoop.com/make-duck-house/

Another option is to buy a pre-made duck or chicken coop from Tractor Supply or Amazon. A simple Google search for "duck coop" will bring up several viable options. However, you must still employ your critical thinking when searching for a model that is suitable for ducks, not only for chickens.

For creative ideas on how to create your duck house and pen, simply check out this thread: https://www.backyardchickens.com/threads/what-does-your-duck-house-look-like.178796/

One person used a child's playhouse, connected to a small dog run, to house their ducks. Another used a chicken tractor, with clever paving stones and rocked-in areas around their duck water bowl and pond. Still others used a simple A-frame house and mesh yard that can be easily moved around to different spots in their forested backyard.

Additionally, you can consider buying detailed plans that will help you construct your own duck house. Many online sites offer ready made plans that will guide you every step of the way.

Carolina Coops will build you a duck coop that will last you a lifetime. They will even come to your location and build it on site. This high-quality product contains Douglas fir wood, PVC coated mesh, and other high-quality ingredients.

Questions to consider:

1. What do you have sitting around your backyard that might be able to be converted into a duck cage? Do you have a doghouse, small play house, large, weatherproof box, shed, or barn where the ducks could shelter at night?
2. Do you have money on hand to buy a pre-made duck coop?
3. Are you creative and handy with tools and wood?
4. Do you have time to build yourself a custom duck coop?

Ventilation

As mentioned above, ventilation is a key component of any duck set-up. Ducks not only love swimming, bathing, and playing in water and mud, but they also create a lot of humidity. If you're a duck, you're always creating moisture—even when you're just breathing![236] If left to themselves, ducks can make their homes into a stew of wet, soupy filth. You'll want to avoid this at all costs. Not only do you want to keep the area clean and fresh for your own convenience and comfort, but an overly humid environment can also be unhealthy for ducks.

An overly humid duck coop can be a perfect environment for breeding pests. Terribly gross worms and insect larvae can live in the duck poop, wriggling in the wet bedding and under flaps of linoleum and layers of straw.

Furthermore, very wet bedding can easily freeze in winter, creating a cake of immovable straw and poop that is almost impossible to clean out. Believe me, you do not want to be stuck on a -10 degree day hacking out giant chunks of frozen poop and straw from a duck coop. Ask me how I know.

In summer, overly wet bedding can create a very gross and smelly environment. Cleaning the coop becomes frustrating and nauseating. Further, wet poop and straw can accelerate the production of ammonia. Excessive amounts of ammonia can cause health risks to animals and humans.

Metzer Farms explains, "Ammonia gas is produced by the breakdown of uric acid in poultry droppings by bacteria in the litter. When wet, the ammonia production is accelerated and is especially prominent in coops where there is a high percentage of manure in the bedding."[237]

The more moisture is retained in the bedding, the higher the risk of ammonia collecting in the coop. You recognize ammonia by the pungent, sour smell of urine that greets your nose when you open the coop. In these cases, you need to leave the coop door open to air out the coop. By the time you can smell ammonia, it is definitely a high risk for harming your ducks and chickens.

Metzer Farms explains, "According to the Occupational Health and Safety Administration, people can start to smell ammonia between 5 and 50 parts per million (ppm) depending on how well they can smell. Anything above 25 ppm and the ammonia is now in danger of damaging your bird's health. Therefore, ammonia can be harming your birds and you don't even notice it!"[238]

---

[236] Lisa Steele, "A Guide to Duck Houses: six things to consider when building a home for backyard ducks," *HGTV*, Accessed May 19, 2021, https://www.hgtv.com/outdoors/gardens/animals-and-wildlife/a-guide-to-duck-houses.
[237] "Ammonia Causes and Effects," *Metzer Farms*, Last modified on December 1, 2017, Accessed on May 19, 2021, https://metzerfarms.blogspot.com/2017/12/ammonia-causes-and-affects.html.
[238] "Ammonia Causes and Effects," *Metzer Farms*, Last modified on December 1, 2017, Accessed on May 19, 2021, https://metzerfarms.blogspot.com/2017/12/ammonia-causes-and-affects.html.

Ammonia can cause many serious health problems for ducks, chickens, and humans. Metzer Farms warns about the dangers of ammonia in duck coops: "Ammonia gas is acidic and can cause serious damage not just to your bird's throat, lungs, and eyes, but your own as well. Ammonia has been known to cause blindness, damage to the esophagus, and death via suffocation."[239]

Airing out the coop is good, but it's not enough. It's important to ensure that the coop has ample ventilation at all times. In a duck house, you must include a screened window, a ventilation hole, a large open door, or vents created just for this purpose.[240] Any way to get airflow is important and needed.

You may worry that allowing ventilation is bad for ducks, especially in winter. You may worry that ducks will get too cold, freeze to death, or get frostbite if air is allowed to circulate during the coldest months.

However, ducks are not a susceptible to the cold as you might imagine, and they are not as susceptible as chickens are. A good layer of straw on the bottom of the coop is enough to keep their sensitive feet warm.[241] So you don't need to worry about allowing airflow for ventilation, even in the middle of winter.

Nesting Boxes. Some ducks use nesting boxes, and others lay their eggs wherever suits their fancy. However, our ducks were reared alongside chickens and developed good habits of laying their eggs in nesting boxes.

However, if nesting boxes are not available, ducks won't complain. They'll lay their eggs anywhere that suits them: on the ground, in a pile of straw, in a pond, or in a dirt hollow.

If you want to provide nesting boxes for your ducks, consider something simple like a cardboard box full of straw. Make sure there is an easy access ramp or lip for your duck to enter the box.

Unique Types of Duck Coops

A duck house can be very simple, because ducks are not picky. However, some duck owners choose to decorate their duck coops and make them fun. How about a beautiful

[239] "Ammonia Causes and Effects," *Metzer Farms,* Last modified on December 1, 2017, Accessed on May 19, 2021, https://metzerfarms.blogspot.com/2017/12/ammonia-causes-and-affects.html.

[240] Lisa Steele, "A Guide to Duck Houses: Here's six things to consider when building a home for backyard ducks," *HGTV,* Accessed May 19, 2021, https://www.hgtv.com/outdoors/gardens/animals-and-wildlife/a-guide-to-duck-houses.

[241] Lisa Steele, "A Guide to Duck Shelters for Winter: Information about Ducks to Keep them Warm and Healthy Through the Winter," *Backyard Poultry,* Last modified on October 26, 2020, Accessed May 19, 2021, https://backyardpoultry.iamcountryside.com/coops/duck-shelters-for-winter/.

little blue house with wooden shingles on the roof and fake flowers in the flower box by the front window? This little house was set by a lake and the ducks enjoyed their elegant habitat.[242]

How about a little house with beautiful wooden siding, that looks like a camper? Place it by a pond with rocks and sand around it, and you'll be good to go.[243]

Others decorate their coop area with beautiful potted and hanging plants, fresh roses, little pink tubs, and matching slides.

How about building your drakes and hens a "Duck House Hotel"? The Morning Chores provides free plans about how to turn your duck house into a creative, beautiful home complete with painted curtains and a sign.[244]

You might want to make a Cobb house for your ducks. It looks like it's made of adobe, with a foundation of real rocks. Complete with wooden doors and a corrugated metal roof, this house will look like your ducks are settlers in an old Anasazi encampment.[245]

In addition, you can paint your duck house bright colors, like a delicate dwelling in the English countryside.[246]

How about a "Gingerbread Duck House?" This fascinating, sweet little place will make your ducks fall in love.[247]

Do you like gardening? You can incorporate planter gardens into the roof of your duck or chicken house. Try this amazing model from Tyrant Farms. You will be charmed by this beautiful little home for your quackers.[248] You can grow fresh sprouts, plants, vegetables,

---

[242] "What Does Your Duck House Look Like," *Backyard Chickens,* Last modified on March 3, 2010, Accessed on May 19, 2021, https://www.backyardchickens.com/threads/what-does-your-duck-house-look-like.178796/page-2.

[243] "What Does Your Duck House Look Like," *Backyard Chickens,* Last modified on March 3, 2010, Accessed on May 19, 2021, https://www.backyardchickens.com/threads/what-does-your-duck-house-look-like.178796/page-3.

[244] Jennifer Poindexter, "37 Free DIY Duck House / Coop Plans & Ideas that You Can Easily Build," *Morning Chores,* Accessed May 19, 2021, https://morningchores.com/duck-house-plans/.

[245] "37 Free DIY Duck House / Coop Plans & Ideas that You Can Easily Build"

[246] "37 Free DIY Duck House / Coop Plans & Ideas that You Can Easily Build"

[247] "37 Free DIY Duck House / Coop Plans & Ideas that You Can Easily Build"

[248] "INTRODUCING THE "QUACKER BOX" DUCK HOUSE – A BIRTHDAY PRESENT WORTHY OF THE TYRANT," Accessed May 19, 2021, Last modified on September 27, 2018, https://www.tyrantfarms.com/introducing-the-quaker-box-a-birthday-present-worthy-of-the-tyrant/.

or fruits on the roof for your ducks, and provide them with a continuous, fresh supply of healthy greens.

Chapter 6: Bringing Your Duck Home

Now that you have prepared a safe and predator-free area for your ducks to live, you are ready to bring them home.

As you purchase your duck, it's important to know how to analyze the health of your bird.

How do I know if the bird is healthy?

Obviously, if you are purchasing from an online breeder, you will not be able to inspect the ducklings personally and find out whether or not they are healthy. If you choose a more personal option, you can look the ducklings over to make sure you are buying the best of the best.

Here are some things to look for when inspecting a new bird:

1. Feet health. Bumblefoot is a common disease in ducks, so please ensure that the duck does not have any lumps or growths on their legs.[249] A duck with bumblefoot will present unique challenges. You will need to clean and lance the foot, soak it in salt water, and perform a number of other somewhat distasteful treatments. It is better to avoid this problem at the outset. A duck's feet should be supple and flexible, and should not have sores, lumps, or injuries.[250] Check to make sure the duck's feet are not swollen.[251]
2. Legs. Legs should have good range of motion. Legs should not make any unusual sounds when in motion. They should be flexible and easily moved. Look for sores and swollen body parts.[252]
3. Feathers. Feathers should be clean and smooth. Ducks' feathers should not be dirty for long, as ducks should be able to clean themselves. Water should roll off them "like water off a duck's back." In other words, when their oil glands are working properly, water balls together and rolls right off. Ducks who look like they are sopping wet after a swim do not have properly maintained feathers. In addition, ducks should not have visible mud or dirt over their wings, backs, and chests (unless they have been recently playing

---

[249] "How to Conduct a Duck Health Examination," Last modified on April 10, 2020, Accessed on May 6, 2021, https://opensanctuary.org/article/how-to-conduct-a-duck-health-examination/.
[250] Ibid.
[251] "How to Conduct a Duck Health Examination," Last modified on April 10, 2020, Accessed on May 6, 2021, https://opensanctuary.org/article/how-to-conduct-a-duck-health-examination/.
[252] Ibid

in the mud). Open Sanctuary explains that duck "feathers should not be dirty, dull, missing, tattered, frayed, ruffled, or broken."[253]

4. Eyes. Examine the eyes for clarity and cleanliness.[254] A duck's eyes should not be "cloudy, watery, dry, swollen, or crusty."[255] Do not purchase a duck that has pus or other secretions in its eyes and nostrils.[256] A duck's nictitating membrane should not be discolored and should be able to open and close over the eye at will.[257]

5. Skin. Clean, pink skin is a sign of health.[258]

6. Chest feathers. A duck should have clear skin on the chest, and a breastbone which is easy to find and not covered with fat. Neither should it be too easy to find, indicating that the duck has not been eating enough.[259]

7. Abdomen. The abdomen should not be too hard or too squishy. A nice, firm softness is a sign of health. Otherwise, the duck could be at risk for "egg binding, yolk peritonitis, egg yolk impaction, a bacterial infection like salpingitis, fluid blockage, or heart failure."[260]

8. Oil gland. The oil gland lubricates the duck's feathers and needs to be in good working condition for the duck to be healthy. Ensure that the preen gland is a normal size. Lumps anywhere on the duck's body are a bad sign.[261]

9. Vent. The vent is the opening of the duck's rear where the egg is laid. This area should not contain blood, poop, dirt, or accumulated crustiness. Furthermore, it should not have infestations of small mites or other unwelcome visitors. Ensure the vent looks healthy and damp.[262]

10. Crop. Birds have a crop which aids in digestion. Similar to other body parts, the crop should not be too firm and knotty. Ensure that the crop is supple and can easily expand to hold food and water.

11. Head. The position of the head is a good indicator of the health and well-being of the duck. Ensure that the head is held confidently and is not trembling or weak.[263]

---

[253] "How to Conduct a Duck Health Examination," Last modified on April 10, 2020, Accessed on May 6, 2021, https://opensanctuary.org/article/how-to-conduct-a-duck-health-examination.

[254] "How to Conduct a Duck Health Examination"

[255] Ibid.

[256] "How to Conduct a Duck Health Examination"

[257] Ibid

[258] "How to Conduct a Duck Health Examination"

[259] Ibid

[260] Ibid

[261] "How to Conduct a Duck Health Examination"

[262] "How to Conduct a Duck Health Examination"

[263] Ibid

12. Bill. The beak is another place that is important to check when assessing a bird's health. Like other parts of the body, the duck's nostril should not have discharge. Bubbling nostrils or nostrils that are misshapen and large may be a sign of sickness.[264]
13. Mouth. Inside the mouth lie other clues to duck health. The duck's "breathing should not be labored, loud, wheezy, rattly, sneezy, whistling, or squeaky."[265] Again, ducks' mouths should not be bubbly or secreting unfamiliar substances. Bad breath is another sign of illness.[266]
14. Wings. Wings should be able to move freely. Similar to other areas of the body, they should be free from wounds and swollen abscesses.[267]
15. Droppings. Evaluating duck droppings is not the most pleasant part of evaluating a duck's health, but it can be important. Here are some signs of sickness in ducks: poop that is "poorly formed, pasty, watery, strong smelling, black, bloody, yellow, neon green, or foamy"[268] can indicate a problem that needs to be addressed.
16. Other causes for concern:
    a. A timid duck that hides and stays away from other ducks and chickens can be a source of concern. Consider taking this animal to the veterinarian.[269]
    b. A duck that opens his or her mouth too much may be sick.[270]
    c. Most ducks spend time with other flock members, so a duck that stays away from other members of their group may not be feeling well.[271]
    d. Most ducks and chickens have an established place in the pecking order, but a sick duck may not be able to stand up to other flock members. Pay attention to animals that seem to be unable to defend themselves against pecking and maltreatment.[272]
    e. Limping is always a cause of concern.[273]

---

[264] "How to Conduct a Duck Health Examination," Last modified on April 10, 2020, Accessed on May 6, 2021, https://opensanctuary.org/article/how-to-conduct-a-duck-health-examination/.
[265] Ibid
[266] "How to Conduct a Duck Health Examination"
[267] Ibid
[268] Ibid
[269] "How to Conduct a Duck Health Examination"
[270] Ibid
[271] "How to Conduct a Duck Health Examination"
[272] "How to Conduct a Duck Health Examination"
[273] Ibid

f. A duck that loses its appetite or drinks too much water may be sick.[274] Any kind of poultry that drinks nonstop may be close to death.

g. Most ducks are very alert, so closing their eyes is a sign of sickness.[275]

h. Ducks love to run around, explore their environment, and eat, so a duck who lays down too much may be sick. In general, look for "immobility, inactivity, or unresponsiveness to your approach" as signs of sickness in ducks.[276]

Enjoying your new baby ducklings. By this point, you may feel overwhelmed. You've read dozens of facts about duck breeds and selected the breed that is best for your family and situation. You've investigated breeders and pored over dozens of details about hatcheries. You've built a secure coop and worried about drainage, predators, and mud. You have carefully inspected your new ducks or ducklings to ensure that they are healthy and strong.

By this time, your head is likely spinning. You are on brain overload. There are so many factors to keep into consideration.

Don't let the details scare you. If you have made it this far, you are a devoted, loyal, and capable duck owner. Once you hold the precious ducklings in your arms, the hard work will be more than worth it. They will win your affection and steal your heart!

The moment has come to buy your sweet ducklings. Don't forget to treasure every moment that they are small. Hold them close to your heart. They won't be little for long, so spend every minute you can with them. You've worked hard, and now you deserve a break. Just kick your feet back, relax, and enjoy the new additions to your family!

Taming your ducklings. If you hatched your own ducklings or bought them from a local breeder where they had been handled, the ducks may be easy to tame. The more you hold them, the tamer they will be. You may enjoy petting them on your lap while you watch TV. Perhaps you want to let them out in the backyard and watch them swim. Some people have their ducklings swim in the bath tub every day. Others buy toys for their ducklings and watch them play. If you want your ducks to be well socialized, be sure to spend plenty of time with them. Again, selecting a calm and loyal breed is key to creating a special duck pet.

---

[274] "How to Conduct a Duck Health Examination," Last modified on April 10, 2020, Accessed on May 6, 2021, https://opensanctuary.org/article/how-to-conduct-a-duck-health-examination/.

[275] Ibid

[276] "How to Conduct a Duck Health Examination"

If you have bought ducklings, make sure you keep them in a warm, safe place.

Chapter 7: Living with Ducks: Daily, Weekly, and Yearly Routines

After you select the perfect breed, evaluate your duck's health, purchase your new best friend, and bring it home, it's time to become acquainted with the daily routine you will need to follow. It will take some time to adjust to the rhythm of raising ducks, but the more you practice it, the more second nature it will become.

Some of your ducks' needs must be attended to daily; others might need help weekly; and still others will become part of your yearly rhythm as seasons change.

Daily Routines

Some of your ducks' needs are daily and repetitive. Every day, you will need to pay attention first to the food and water needs of your new pets.

Water

One of the most important elements of your duck's life is its water consumption. We all know that ducks love water, so it is important to not only provide drinking water, but also a place to play and swim.

Ducks need to rinse their nares and eyes in fresh, clean water every day. They need to be able to submerge their entire heads, so don't just give them a chicken waterer. They need a bucket or pail with sufficient space for cleaning.

It's not absolutely necessary to provide a pool for swimming. Some ducks are fine with using a bucket or tub. Sometimes, they will climb in and splash around. Other times, they will stand near the bucket or tub and clean themselves by splashing water all over their feathers. This is why it's important to take care of mud management.

Ducks also love using water to wash down their food. If they found some particularly rich and yummy mud with insects and other critters inside it, they will dunk the mud in the water to rinse it and help them filter out the yummy snacks they've discovered. It's important to provide water for them to drink while they eat. They like to take a drink after almost every mouthful. Sometimes, they put their food in the water before they eat it.

Types of water bowls and pools

Water bowls and pools need to be calculated into the equation before you bring your duck home. Logistical challenges, watering problems, and other obstacles need to be factored into the equation.

Let's talk about some of the different water contraptions you may want to consider adding to your duck enclosure. These are optional, and don't need to be in place right away. You may start with a simple dog bowl, and later move to a more elegant pond contraption.

Here are some of the most common duck water options:

Dog bowl. A large dog bowl works fine for most purposes. Make sure it's big enough for the ducks to submerge their head in it.

Horse bucket. A simple horse bucket will provide plenty of space for your ducks to submerge their heads. Hang it on a low nail or secure it to the wire mesh to ensure that it does not tip over.

Stock tank. If you have access to a stock tank, your ducks may be thrilled to have a large area to swim in. Ensure that your stock tank has a ramp both in and out of the pool, so your ducks have easy access and a way to get out. Some ducks are easily terrified if they don't feel like they can get in and out of the water. If there is not an easy way out of the pond, some ducks may never go into a pond again.

In addition, you can easily deice a stock tank by buying a submersible stock tank deicer. The large volume of water in the stock tank means you will have to change it out less often than a smaller water container.

Kiddie pool. A kiddie pool is an excellent, low-cost option for backyard ducks. Any local dollar store, Walmart, or other general merchandise store should have child sized pools in summertime.

The advantages of kiddie pools are that they are easily available and relatively inexpensive. They are big enough that ducks can swim, often getting their feet off the bottom and paddling around. They are also small enough that ducks can get in and out relatively easily.

The disadvantages of kiddie pools are that they are relatively flimsy. You will discover that you need to empty and clean your kiddie pool almost every day. After a certain number of times of flipping and cleaning them, kiddie pools can crack or buckle. Sometimes, they get holes from sharp rocks or pieces of wood or metal that pierce them. They may not last you more than one year.

To preserve the life of your kiddie pool, try stacking two pools together. This gives more strength to leverage when turning it upside down, and it prevents being easily pierced.

Natural pond. Of course, this is the easiest type of pond. If you have a natural pond or lake on your property or in your backyard, you will let your ducks roam on this natural pond. Consider making a secure house where they can be shut up at night to keep them safe from predators. During the day, the ducks can freely explore their aquatic habitat.

Pond feature. Are you consider building a pond feature in your backyard? Keep in mind that the pond must be self-cleaning, or you will be left with a disastrously messy and mucky pool that will not add any beauty to your backyard or bring any joy to your waterfowl.

An important way to keep a pond feature clean is by adding a filter. There are many types of pond filters on the market. A typical submersible pond filter can do the trick.

You can also buy a biofilter or create your own do-it-yourself biofilter system. Tyrant Farms provides detailed instructions about creating your own, do-it-yourself bio filter system. Check out their instructions and invest with time, labor, and money in building a natural filtering system for your backyard pond.

If you have a large, stagnant pond at risk for growing algae, you may want to consider adding barley straw to your pond. This helps naturally clean away algae.[277]

Self-cleaning do-it-yourself duck pond. Tee Diddly Dee offers instructions to an amazing, self-draining duck pond that you can build in your own backyard. Dig a deep hole, install a duck pond with internal ducks (so the ducks can easily exit the pond), and install a drain that opens into the deep hole below. Check out her instructions for more information: https://www.teediddlydee.com/diy-easy-drain-duck-pond/

The Island Farmer shows how to build an Easy Cleaning Duck Pond with an efficient, do-it-yourself drain. He uses a black plastic duck pond and inserts a drain system made of PVC pipe and PVC cement. You'll have to check out his detailed instructions if you're interested in building your own DIY duck pond with a drain.

You can build a do-it-yourself duck pond filter system with trash cans, lava rock, PVC pipes, scrubber pads from Home Depot, garden rock, and other household items that are easy to access. This model will drain a large amount of water. Check out the instructions here: https://www.youtube.com/watch?v=VMmvKpNrcTg

The water cleaning system may not make your water completely clear. It may remain "tea-colored," but nevertheless, the filter works very nicely to clean your water.

Keep in mind that a DIY duck pond filter system may sound easy, but it is often very time consuming and complicated. Take a look at this list of necessary supplies for a do-it-yourself duck pond filter system from Runner Duck page:

1. 3/4" PVC Pipe (2)

---

[277] Hana LaRock, "How to Keep a Duck Pond Clean Naturally," *SF Gate,* Accessed May 19, 2021, Last modified on October 8, 2019, https://homeguides.sfgate.com/keep-duck-pond-clean-naturally-84449.html.

2. 2" PVC Pipe (1)

3. 1" PVC Pipe (1)

4. 2" 90 Degree Elbows (2)

5. 2" Male Adapter, for 1/2 of bulkhead adapter

6. 2" Female Adapter, for 1/2 of bulkhead adapter

7. 1" Male Adapter (2), for 1/2 of bulkhead adapters

8. 1" Female Adapter (2), for 1/2 of bulkhead adapters

9. 1" 90 Degree Elbows (1)

10. 1" PVC Pipe to Garden Hose Fitting (2)

11. 3/4" 90 Degree Elbows (8), for supports

12. 3/4" Kris Cross Fitting (2), for supports

13. 44 Gallon Rubbermaid Trash Can

14. Egg Crate Louver from overhead fluorescent light fixture, for supports

15. Washable/Reusable Furnace Filters (10)

16. Water Pump sized per the above instructions

17. Small Rubbermaid RoughTote to house the pump

18. 3/4" PVC fittings to get from the pump to a garden hose on the outside of the box."[278]

Building a pool on an incline. A duck pond can be set into the hill so that it is easier to get into. A black plastic pond shell, partially buried in the side of a hill, allows the duck a gravity-driven way to both enter and exit the pond. They can flop into the pond from above, since the pond is set into the hill and does not have a lip. This will make the pond easier for the ducks to access. Furthermore, the ducks can also exit the pond with gravity, simply flopping over the bottom lip at the bottom of the pond.

Another way you can use a hill to your advantage is by making a pond out of pond liner and rocks. I saw an example of a pond that was made simply out of pond lining sheet. The downhill side was blocked off with bricks and large rocks, and the pond lining lay on top of the bricks and large rocks which formed the lower edge of the pond. When it was time to clean the pond, the duck owner simply removed one or two of the bricks or large rocks, allowing the flexible pond liner to bend and the water to run downhill out of the pond. The pond could then be rinsed while the blocks were still out of place. Then, the bricks or large

---

[278] "Do It Yourself Bio Filter," Accessed May 19, 2021, http://www.runnerduck.com/pf1.htm.

rocks can be replaced, the pond liner can be placed back on top of the rocks, and the pond can be refilled.

Mixing tub. One of the best and simplest solutions for duck ponds is a simple concrete mixing tub or concrete mixing pan from Home Depot or Lowes. These simple plastic tubs are often inexpensive, easy to use, and easy to replace. They are strong and sturdy, simple to empty and refill, and easy to rinse. Many duck owners are initially excited about creating giant water features or ample ponds for their ducks. However, in the end, a simple concrete tub will do the trick. It's big enough to swim and splash around in, small enough to fit easily in a small coop, and is a perfect size for daily refilling. In other words, you feel wasteful when you dump an entire kiddie pool once a day. But dumping a concrete tub feels less wasteful. Most concrete tubs hold about five gallons of water, but you can buy them in different sizes. The black plastic tub can retain heat during the winter, making the water less likely to freeze.

Livestock feed bowl. Livestock feed bowls of various sizes also provide ample water sources for ducks.

Providing access for the ducks to the pool. As alluded to above, ducks need a ramp or another way to access the pond. Sometimes, they are comfortable jumping or flying into a raised pond lip, but they can be dissuaded by something as short as a kiddie pool. If your pond is deep or high, consider digging a hole and partially submerging the pool in the ground. Even if you submerge the pond in the ground, the ducks may be discouraged by being unable to get out of the pool. Make sure your pond has steps or ramps to exit the pool as well. Ducks are easily scared and frightened by having to flop their heavy, wet bodies over the lip of a pool that seems too high or too formidable to them.

Keep in mind that ducks must have access to clean water, deep enough to dunk their heads in, at all times. This includes winter, so make sure to buy a heated water bowl to keep your water liquid during freezing temperatures.

Water source. In addition to a water bowl, you will also need a reliable source of water. Some duck owners choose to install a pump, spigot, or faucet directly beside or above the duck water cage. This provides a continual source of water for refilling, rinsing, and cleaning.

Others choose to run a long hose from the house or barn to the duck coop. Personally, I purchased a one-hundred-foot metal hose that has served me well. Since it is metal, it does not run the risk of being punctured by a dog or being pierced or shredded by a lawnmower. Since it's one hundred feet long, it reaches all the way to my duck cage without any problem.

Other duck owners may choose to collect rainwater on top of their cage, which can then be emptied through a spigot or hose into the ducks' enclosure.

Still, other duck owners choose to manually carry water every day from the house or barn to the duck coop. This is, clearly, the most work-intensive and tedious route of all. Carrying gallon after gallon of water to the duck coop, in the dead of winter, is a very difficult task. However, if this is your only option, you may want to invest in a sturdy wagon which you can use to haul the water. Furthermore, you may want to buy large, tough five-gallon containers with spouts, to reduce the number of times you will have to go back and forth between the house and the duck coop.

Daily Water Routine

Changing your ducks' water will likely be a part of your everyday routine. Once you have selected a suitable water dish and water source, you will simply do the simple, daily tasks of dumping, rinsing, and refilling the dish.

Ducks soil their water every day, sometimes within a few minutes or hours. They love putting mud, food, grain and poop in their water, so don't expect the crystal-clear pool of water to last for long. Instead, expect to have to empty and refill your ducks' water bowl frequently.

The constant refilling of water bowls and pools is the reason that drainage is so important. After you empty the water dish, ensure that the water drains away from the coop. Perhaps it drains through a ditch, into a sandy area, or down a drainage hole located directly under the water bowl. Otherwise, your ducks will have a heyday playing in the water and mud. Of course, it's fun to watch them play in the water, but it's better if ducks play in clean water than in dirty water. Too much mud can deteriorate their feathers and cause Wet Feather.

Feeding Your Ducks
Another duck need that requires attention is feeding. Whether you purchase your duck online or through a local store or breeder, purchasing food needs to be one of the first things you think about as you bring your new pet home.

Feeding your ducks may seem simple. Buy some food, put it out. Feed some snacks. Worst case scenario, throw breadcrumbs. Done.

If you have raised chickens, you know that feeding is about that simple, am I right?

However, the feeding and nutrition of ducks is surrounded by several debates. Let's look at these debates one by one.

The debate over medicated and non-medicated food for waterfowl. Traditionally, there has been concern over buying ducks medicated food. Medicated chick food was supplemented with antibiotics to help chicks ward off infections. The particular type of antibiotic used was poisonous to ducks. Duck breeders passed down the important information that medicated feed could poison ducklings.

In more recent years, that myth has been debunked by some sources. Some sources say that feed suppliers now use a different kind of antibiotic that is no longer harmful to waterfowl.[279] K J Theodore explains, "Medicated Starter with Amprolium or Bacitracin in it will NOT harm your waterfowl. I have confirmed this with two Poultry Research Veterinarians."[280] K J Theodore has used medicated starter with both ducklings and chicks, without any harmful effects to either type of bird. He states that the myth of poisonous medicated starter "was actually once true years ago when different types of drugs were used in medicated feeds. But with the use of Amprolium in particular now, it is no longer a concern."[281] A medical study was performed on Khaki Campbell ducklings to learn about the safety of the updated medicated feed. The results were reassuring: "No significant differences in mean body weight, mortality, and anatomical development were observed among the treatments."[282]

What Type of Food? Do you need to use waterfowl feed for your ducks? Is it okay to use chicken layer mash for ducks? Should ducks' diets be supplemented with vitamins and minerals?

Again, different breeders have different opinions on these topics. Some sources say that it is perfectly fine to feed ducks layer mash designed for chickens.[283]

Most sources agree that if you feed your ducks chicken layer mash, it's important to add niacin supplements, such as Brewer's Yeast.

Another supplement high in niacin is raw wheat berries. Many sources say that it is a good idea to feed your ducks wheat.[284] Wheat contains a lot of niacin, which is a critical

[279] K J Theodore, "Common Poultry Myths," *Shagbark Bantams,* Accessed May 20, 2021, https://www.shagbarkbantams.com/common-poultry-myths/.

[280] K J Theodore, "Common Poultry Myths," *Shagbark Bantams,* Accessed May 20, 2021, https://www.shagbarkbantams.com/common-poultry-myths/.

[281] "Common Poultry Myths"

[282] D Holderread, H S Nakaue, G H Arscott, "Anticoccidial drugs and duckling performance to four weeks of age," 1983 Jun; 62(6) : 1125-7. doi: 10.3382 /ps.0621125, Accessed May 20, 2021, https://pubmed.ncbi.nlm.nih.gov/6878147/.

[283] Tim Daniels, "Feeding Ducks," *Poultry Keeper,* Last updated on September 11, 2018, Accessed May 20, 2021, https://poultrykeeper.com/duck-keeping/feeding-ducks/.

[284] Lisa Steele, "Can Chickens & Ducks Share Feed?," *Hobby Farms,* Last modified on January 2, 2017, Accessed on May 20, 2021, https://www.hobbyfarms.com/can-chickens-ducks-share-feed/.

ingredient for ducks. Placing the wheat at the bottom of a tub of water prevents your other poultry from eating it.[285]

In addition, you can also choose to feed your ducks nutritionally balanced waterfowl food. You can buy Manna Pro Duck and Goose Layer Pellets, a complete feed for laying ducks and geese, on Amazon. This feed is "crafted with calcium for strong shells."[286]

No matter what you choose to feed your duck, ensure that they have plenty to eat every day. You may choose to feed your ducks once a day (in the morning), or twice a day (morning and evening). You may choose to leave an unlimited amount of food for your ducks to snack on constantly. You may choose to feed them a smaller amount of grain and mash, and supplement their diet with worms, greens, grass, and weeds in your yard.

Fermented food, fodder, and sprouts. In today's world, it's popular to feed your ducks fermented food, fodder, and sprouts.

1. Sprouts. In order to create healthy, fresh, green vegetables for your ducks, soak wheat berries in water overnight. After soaking, drain the wheat berries. Keep them moist by rinsing them every day. Soon, the wheat berries will begin to grow. The fresh sprouts will delight your ducks.
2. Fodder. To create fodder, allow your sprouts to grow a little longer, until you see a lush carpet of newly grown wheat. It will look like grass. Give the fodder to your ducks. They will be delighted. Fodder benefits your ducks by providing extra nutrients, especially during winter. When there is no fresh grass growing outside, your ducks will thank you for providing delicious, green wheat grass for them to eat.[287]
3. Fermenting your duck food helps break down the nutrients in the food, releasing them more readily to the ducks' digestive systems.[288] Additionally, the Fresh Eggs Daily blog informs us that "Fermentation introduces vitamins, specifically the B vitamins (folic acid, riboflavin, niacin, and thiamin), not present before fermentation, so if you are

[285] Lisa Steele, "Can Chickens & Ducks Share Feed?," *Hobby Farms,* Last modified on January 2, 2017, Accessed on May 20, 2021, https://www.hobbyfarms.com/can-chickens-ducks-share-feed/.
[286] "Manna Pro Duck Layer Pellet | High Protein for Increased Egg Production | Formulated with Probiotics to Support Healthy Digestion | 8 Pounds," Amazon description, Accessed May 20, 2021, https://www.amazon.com/Manna-Pro-Duck-Layer-Pellets/dp/B0793FPLMR.
[287] "Growing Fodder for Backyard Chickens, Ducks, and Geese," *Fresh Eggs Daily,* Accessed May 21, 2021, https://www.fresheggsdaily.blog/2014/02/growing-sprouted-fodder-for-your.html.
[288] "How to Save Money by Fermenting Chicken Feed," *Fresh Eggs Daily,* Accessed May 21, 2021, https://www.fresheggsdaily.blog/2015/03/how-to-save-money-fermenting.html.

currently adding brewer's yeast to the feed, you don't need to any longer."[289] For duck owners who are perplexed about adding Brewer's Yeast, this is an instant answer to your questions! Voila! You will no longer have to hunt down this rare type of yeast to add to your duck mash. Nor will you have to worry about your ducks' nutritional deficits, which might cause disease or Wet Feather. Fermented food actually strengthens your ducks' immune systems, helping them fight off disease. In order to create fermented duck feed, "just add water." No, really. All you need to do is add water and stir. The amazing properties of water and grain does the rest. The first day, you must soak the mash, pellets, or grain in water.[290] The water should not contain chlorine, since chlorine will kill the good bacteria that you want to be present in the fermented food. You can buy distilled water at the store, or you can allow your water to sit out overnight to allow the chlorine to escape.[291] Every day after that, you must stir the fermenting grain and add enough water to cover the grain.[292] That's all there is to it! In three days, the grain will smell sweet and mildly fermented. It's time to feed it to your ducks! You will save money on feed through using fermented feed. Because fermenting releases nutrients that were stored inside the grain and were previously hard to access, your ducks will get more nutrients per pound than they were before, and they will eat less.[293] Be aware that fermented food will be a bit messy when you add ducks to the equation. The ducks will slurp and fling it around everywhere! However, you may find that fermenting your duck food is worth every bit of mess. Furthermore, be aware of things that could go wrong with fermenting. If part of the mash is not covered with water, it runs the risk of becoming truly putrid. If your fermented feed smells gross, sour, or yucky, it is most likely rotting, rather than fermenting. Dump it out and start over.

Nutritional Needs of Ducks

Ducks need 55 mg of niacin in their daily feed.[294] Additionally, they need 60 mg per kg of zinc, 80 mg per kg of iron, 50 mg per kg of manganese, 8 mg per kg of copper, and a small

[289] "How to Save Money by Fermenting Chicken Feed," *Fresh Eggs Daily,* Accessed May 21, 2021, https://www.fresheggsdaily.blog/2015/03/how-to-save-money-fermenting.html.

[290] "How to Save Money by Fermenting Chicken Feed"

[291] "How to Save Money by Fermenting Chicken Feed"

[292] "How to Save Money by Fermenting Chicken Feed"

[293] "How to Save Money by Fermenting Chicken Feed"

[294] "Nutritional Requirements For Ducks and Geese," *Metzer Farms,* Accessed May 21, 2021, https://www.metzerfarms.com/NutritionalRequirements.cfm.

amount (less than half a milligram per kilogram) of iodine and selenium.[295] Furthermore, ducks need amino acids: arginine, histidine, isoleucine, leucine, lysine, TSAA (Met+Cys), Threonine, Tryptophan, and Valine.[296] They need 15 KIU per kg of Vitamin A, 3 KIU per kg of Vitamin D, 20 IU per kg of Vitamin E, 1 mg per kg of Vitamin K, 0.2 mg per kg of Biotin, 1000 mg per kg of Choline, 1 mg per kg of folic acid, and 11 mg per kg of pantothenic acid.[297] In addition, ducks need 2 mg per kg of Thiamine, which is also known as Vitamin B 1.[298] They need 4 mg per kg of riboflavin, which is also known as Vitamin B 2.[299] Ducks need 3 mg per kg of Pyridoxine, which is also known as Vitamin B 6.[300] Finally, they need a very small amount of Vitamin B 12: 0.01 mg per kg.[301]

In other words, ducks have high nutrition needs, specifically adjusted for their unique physical needs. As discussed above, you can meet these nutritional needs through buying a special waterfowl mix, or you can feed them chicken food with special nutritive additives.

Can I feed ducks and chickens side by side?

As discussed above, ducks can be fed layer mash formulated for chickens. Sometimes, however, ducks struggle from nutritional deficits when they are restricted solely to chicken food. If feeding both types of poultry at the same time, consider buying a type of food formulated especially for mixed poultry. For example, Manna All Flock Crumbles is a "complete feed for chickens, ducks, geese, turkeys, and gamebirds."[302] It contains "16% protein, probiotics to support digestion," and is in "crumbled form for easy feeding."

Feed bowl. You can feed your ducks in almost any receptacle, tray, bowl, or platter. Ducks are not picky about feeders. They would love to "inhale" food out of any kind of bowl imaginable. You can throw grain on the ground, put it in a simple dog bowl, or submerge grain in the ducks' water bowl.

Ducks eat a lot. Some people advocate feeding ducks unlimited food, and others prefer to limit the intake of their ducks so that they don't run up too high of a feed bill.

---

[295] "Nutritional Requirements For Ducks and Geese" *Metzer Farms,* Accessed May 21, 2021, https://www.metzerfarms.com/NutritionalRequirements.cfm

[296] "Nutritional Requirements For Ducks and Geese"

[297] "Nutritional Requirements For Ducks and Geese"

[298] "Nutritional Requirements For Ducks and Geese"

[299] "Nutritional Requirements For Ducks and Geese"

[300] "Nutritional Requirements For Ducks and Geese"

[301] "Nutritional Requirements For Ducks and Geese"

[302] "Manna Pro All Flock Crumbles | 16% Protein Level | Complete Feed for Chickens, Ducks, Geese, Turkeys and Gamebirds | Probiotics to Support Digestion | Crumbled Form for Easy Feeding | 25 Pounds," Amazon, Accessed May 20, 2021, https://www.amazon.com/Manna-Pro-Probiotics-Formulated-Vitamins/dp/B01N4KD9MY

Additionally, if your ducks forage in the backyard, they will not need as much additional food. However, it's still important to supplement your ducks' diet with nutritious grain or mash.

Daily Routine

Once you've bought the food, set up a food bowl, and procured special additives or nutrients, it's time for feeding! As you develop your daily rhythm, you will discover what works best for you and your ducks. My ducks start quacking for their food early in the morning. Sometimes, it's as early as 4 am! For this reason, you will most likely not want to wait too long to feed them in the morning. Bright and early, pour their feed into their bowls, and let them eat. You may want to give them another snack in the evening. Ducks do not need to have feed available at night. If you let them out in the backyard to forage for greens and insects during the day, you may put them into their cage at night and let them sleep in peace without making a mess of food and water!

Favorite Foods and Table Scraps

If you spend any amount of time on the popular Duck Facebook group: https://www.facebook.com/groups/backyardducks you will discover an important truth: DUCKS LOVE PEAS!!!

Ducks are almost as good as pigs for devouring any type of table scrap. They are a good supplement to your recycling project. They will eat all types of food waste and convert it into manure for your compost pile.

Ducks love peas, mealworms, and almost anything. They love papaya, watermelon seeds and rinds, cantaloupe seeds and rinds, chicken, pork, beef, octopus, shrimp, and all other types of meat, apples, grapes, beans, rice, wheat, and basically any kind of table scrap you can imagine.

Since ducks are so eager to eat, and so indiscriminate about what they devour, it's important that you, as the duck owner, pay attention to what you put in their bodies. They will eat things that make them sick, such as hot pepper, avocados, or white bread.

Here are a few types of scraps to avoid when feeding your ducks. You want to do all you can to keep your birds healthy, energetic, and happy!

1. "Bread, chips, crackers, donuts, cereal, popcorn, and similar bread-type products or junk food."[303]
2. Potato[304]

---

[303] Melissa Mayntz, "What do Ducks and Other Wildfowl Eat?," Last modified on December 23, 2020, Accessed May 20, 2021, https://www.thespruce.com/what-to-feed-ducks-386584.
[304] Kayla Lobermeier, "What not to feed chickens and ducks!" *Under a Tin Roof,* Accessed May 20, 2021,

3. Tomato leaves[305]
4. Eggplant leaves[306]
5. Avocado[307]
6. Apple seeds[308]
7. Rhubarb[309]
8. Raw Beans[310]
9. Onions[311]
10. Chocolate[312]
11. Coffee[313]
12. Too much spinach[314]
13. Too much iceberg lettuce[315]

Food Storage

Mice and rats love snacking on grain, layer mash, and any other kind of duck goodies you may have on hand. In order to keep the mice and other rodents at bay, it's important to store duck food in predator proof containers. One of the most important features of predator proof containers is their lids and their locks. Raccoons and other creatures are incredibly smart and have agile fingers that can open locks. If you keep your food in a shed where raccoons can sneak in, make sure that your containers are completely immune to the attacks of these curious creatures. Furthermore, it's important that your food storage containers be made of metal or some other tough, durable substance. Mice and other rodents can chew holes in what seems to be very sturdy plastic. Of course, storing your feed in the bags it came in is not a good idea. Mice can chew right through the plastic bag. You don't want to wake up in the morning with half your feed bag spilling out on the ground, pouring out through a hole chewed by a mouse during the night.

---

https://www.underatinroof.com/blog/2018/6/14/what-not-to-feed-chickens-and-ducks.
[305] "What not to feed chickens and ducks!"
[306] "What not to feed chickens and ducks!"
[307] "What not to feed chickens and ducks!"
[308] "What not to feed chickens and ducks!"
[309] "What not to feed chickens and ducks!"
[310] "What not to feed chickens and ducks!"
[311] "What not to feed chickens and ducks!"
[312] "What not to feed chickens and ducks!"
[313] Kayla Lobermeier, "What not to feed chickens and ducks!" *Under a Tin Roof,* Accessed May 20, 2021, https://www.underatinroof.com/blog/2018/6/14/what-not-to-feed-chickens-and-ducks.
[314] "What not to feed chickens and ducks!"
[315] "What not to feed chickens and ducks!"

Letting them out in the morning, putting them in at night.

Another part of a duck's daily routine is being let out of the coop in the morning and being put back into the coop at night. Some people leave their ducks inside their enclosed run all day long. It's certainly the most secure option if you have a lot of daytime predators around. But if you're comfortable with letting your ducks forage for insects and weeds, you may want to allow them to roam freely around your farm or backyard during the day. Most ducks are overjoyed to be allowed to escape their pen every morning.

Just as important as letting them out in the morning, it's important to put ducks in their coop at night. Ducks do not have natural defenses against predators. Most of them cannot fly, and they are an easy snack for marauding enemies. For this reason, it's important to put them in a safe place at night.[316]

Eggs

Many varieties of ducks lay large, pearly, white eggs. Others lay green, gray, or brown eggs. As mentioned above, ducks may lay their eggs anywhere throughout the coop or yard. Others prefer to nest in a specific place every day. My ducks love laying their eggs in the nesting boxes, where they are easy to find. When the ducks are out in the yard, they make themselves a little nest out of straw, leaves, and dirt, and consistently lay their eggs there every day.

Gathering duck eggs. An important part of your daily routine is gathering eggs. If your ducks have the good habit of laying them in the same place daily, this routine is as simple as picking them up. If the eggs are scattered around the yard or coop, you may have to go on an Easter Egg hunt! You may find that your dog has carried them away or even eaten them! Or perhaps they've laid their eggs in the most inconvenient place possible: in the pond, under a table by a mud pit, or somewhere else. It's as if they wanted to watch you crawl on your hands and knees through the mud, wade into the water, and generally make a fool of yourself!

Cleaning duck eggs. Many sources agree that eggs are covered with a natural antibacterial coating. If you keep your duck pen clean and free from mud, your eggs will be generally clean as well. They can be stored without washing, as washing removes the

---

[316] Lisa Steele, A Guide to Duck Shelters for Winter: *Information about Ducks to Keep them Warm and Healthy Through the Winter," Backyard Poultry,* Last modified on October 26, 2020, Accessed May 21, 2021, https://backyardpoultry.iamcountryside.com/coops/duck-shelters-for-winter/.

antibacterial coating. This is why some people store unwashed, natural duck eggs on the counter for some time; they are naturally protected.

Once you wash them, however, the antibacterial film is removed.[317] Then, they should be stored in the refrigerator to keep them free from contaminants.

I've noticed that my duck eggs have a clean, dry surface. I can easily write on the surface with a pencil. However, once I've washed the duck egg, it becomes somewhat slimy. It's very difficult to write on with a pencil after washing.

If you want to avoid removing the antibacterial film (called a "bloom") from the egg, but it is dirty with poop, you can brush the poop off with a dry cloth or a toothbrush.[318]

Storing eggs for food. I store them in the refrigerator. I have habit of using a pencil to write on each egg the date I collected the egg. This helps me make sure I use the oldest eggs first, and always keep them rotating through in the order they were collected.

Hatching eggs. If you plan to hatch eggs, they should not be washed or refrigerated.

Egg cartons. If you've spent any amount of time with duck eggs, you have probably realized that duck eggs cannot be stored easily in normal chicken egg cartons! Duck eggs are giant, and it's difficult to get them to fit! After a few weeks or months of frustration, you'll want to make sure you buy your own duck egg cartons, which are custom made for giant duck eggs!

Weekly Routines

In addition to the daily routines, you will also find that duck care involves weekly routines, such as cleaning the coop.

Some people choose to clean their coop daily. They find it necessary and convenient to rake poop out of the sand substrate every day. Others, who are using the deep litter method, add straw and mix their deep composting bedding every day.

However, my goal with raising ducks was to find tips and tricks to reduce the frequency of the coop cleaning maintenance job to once a week.

---

[317] Phil, "Should You Wash Eggs Before Cracking Them?" Accessed May 21, 2021, https://chickenandchicksinfo.com/should-you-wash-eggs-before-cracking-them/.
[318] Phil, "Should You Wash Eggs Before Cracking Them?" Accessed May 21, 2021, https://chickenandchicksinfo.com/should-you-wash-eggs-before-cracking-them/.

Before we discuss cleaning the coop, let's look at a few tips that will make the coop easier to clean in the first place.

Managing Mud

First, it's crucial to learn to manage mud. Too much mud negatively affects your mental health, and also negatively affects the ducks' physical health.

One of the first and most important ways to manage mud is to put a roof over the coop. As discussed above, you can make a simple roof out of a tarp, a sheet of metal, or pieces of polycarbonate. Using a roof greatly reduces the amount of mud in the coop.

Another method of controlling mud is using the right substrate. As discussed above, you may want to choose sand or another easily drained substrate to help reduce mud.

If you have plenty of cash on hand and would like to invest it in mud reduction, consider purchasing the Lighthoof Mud Management system. Lighthoof will send you a free sample of their honeycomb-like grid. Simply place the grid on the ground and fill it with your sand or other substrate. This helps reduce the amount that the sand is ground into the dirt below. This greatly reduces the amount of mud in your coop.

Bedding, Bedding, and More Bedding

Remember the description about "poop icing"? Yeah. That's the horror that I'm trying to save you from in this chapter.

As I'll mentioned earlier, I started out raising my ducks and chickens together. Later on, I separated my ducks from my rooster because he got unexpectedly aggressive toward them. For a while, I said to myself, "Why not keep the ducks completely separated from the chickens? That way the chickens won't constantly scratch and fling dirt and straw into the ducks' water, and it will stay cleaner." HA. The water may have stayed cleaner, but not the bedding. In only a few days, the straw became coated with a slimy, sticky substance. Underneath, the straw and leaves may have been perfectly dry and fine. But without chickens to turn and rotate the compost, it became a disaster zone. I quickly learned not to ever keep ducks by themselves. Instead, I keep plenty of chickens with the ducks. Even though the chickens may dirty the water with their scratching, it's worth it. If I must keep the ducks by themselves for some amount of time, I try to always rotate the ducks and chickens regularly, so the ducks don't stay by themselves in any particular part of the pen for long.

Different types of bedding:

- Pine pellet bedding. By far the best bedding for indoors is pine pellets. These pellets may be sold as rabbit bedding, horse bedding, or combustibles for a wood stove. If you are keeping ducks inside the coop for an extended period during extreme cold, you will need to put down pine bedding under the straw. Pine chips are also excellent for putting at the bottom of nesting boxes. Ducks tend to bury their eggs, and they also bury their poop. You may see a nest full of perfectly clean straw, only to discover that underneath, there is an infestation of manure, broken eggs, and yuck. Putting pine pellet bedding at the bottom of the nesting box helps mitigate this disaster. In addition, check regularly for eggs and don't be deceived by a beautiful external look. Pine pellet bedding disintegrates when wet and dries quickly. Rather than becoming sticky, caked, or frozen, pine pellet bedding helps the poop remain crumbly. This makes it easier to scoop out when you're ready to clean.

- Straw. Straw is another excellent resource for bedding. This bedding can be placed both inside the shelter and outside in the yard. Straw helps keep down mud when placed on top of sand or another substrate. Continue to add new straw so that the straw does not become matted. Keep in mind that straw on top of mud will simply create what some people refer to as "concrete." This mud-poop-straw cake is no fun to remove later. So, make sure the straw is indoors, or on top of sand or another easily drainable substrate.

- Hay. Hay is another option that can be used both indoors or outdoors.

- Dry leaves. In fall, I often substitute dry leaves for straw or hay. They work great both inside the duck house and outside in the run. They will decompose and become a part of your natural compost.

- Pine shavings. Pine shavings are another favorite of mine. They smell great, absorb a lot of moisture, and don't stick together when wet and poopy. I highly recommend pine shavings for indoors! They could also work well outdoors, although they are quite expensive per square foot, so I don't use them outdoors.

- Koop-Clean. This unique bedding option comes complete with a special additive that keeps down odor! According to the website, "KOOP CLEAN offers a unique blend of chopped hays and straw, combined with the superior odor neutralizing ingredient, Sweet PDZ™, leaving you with a happy flock and a fresh, dry coop all year round."[319]

---

[319] "KOOP CLEAN™ CHICKEN BEDDING: Superior Chicken Bedding for a Fresh, Happy Flock" *Lucerne Farms,* Accessed May 21, 2021, https://lucernefarms.com/koop-clean/.

Other great options for bedding are chopped straw, Aspen shaving nesting liners, and premium pine shavings.

To save time and energy, use the deep litter method. Simply continue to add fresh straw on top of your previous straw inside the house. Add grain to the straw so that your chicken inhabitants will scratch up the straw. Eventually, the poop and straw will become a rich, deep compost.

Cleaning the Coop

Once all these prerequisites are in place, cleaning the coop should be "easy"! Simply sweep out old bedding and replace it. If you are using deep litter, you will simply need to add new straw to the top of your old straw! Rake the sand outdoors, or simply add new straw on top of it! You will find that a lot of investment and effort up front will save you a lot of time and energy later on!

Yearly Routines: Getting Your Ducks Through the Winter

One of the most important parts of the yearly routine is preparing for the winter. It's part of the natural rhythm you'll develop every year with your ducks. When the leaves start falling from the trees, it's time to think about winter proofing and weather proofing your cage.

You'll want to make sure you have equipment to get you through the winter. One of the most important elements of getting through the winter is heated water bowls. It's critical to have an anti-freeze heated water bowl. This can be as simple as a heated dog bowl, or as complex as a submersible stock tank heater.

There are other options that do not include electricity. The higher the volume of water you have in the container, the less quickly it freezes. Furthermore, black absorbs sunshine, so a large, black rubber tub will retain some of the sun's heat. In addition, you can insulate your black rubber bowl. Simply place a rubber cattle feed bowl inside a tire. Fill the inside of the tire with insulation, such as leaves, plastic bags, or other type of insulating material. Then set it in the sunshine. The black tire and the black bowl, along with the insulation, will keep it from freezing as quickly. You can put an A-frame of old glass windows over the contraption. Leave room for the ducks and chickens to drink from one side of the A-frame. Cover the other three sides with glass to create a greenhouse effect.

Inside the chicken house, ensure that there is plenty of straw. Make sure that your shelter is ventilated, but otherwise warm and safe. Ensure the roof is not leaking and that your ducks will stay dry. However, don't over-stress about their warmth. Ducks are hardy animals that don't mind the cold.

Chapter 8: Hatching and Incubation: Raising a New Crop of Ducklings

Incubation Basics

Incubating eggs is a complicated process that takes much time and attention. To become fully equipped to incubate eggs, you will need to do a significant amount of research. Most importantly, read the instruction manual from the incubator of your choice.

Here are a few of the most important things you will need to know about hatching eggs.

First, you must obtain fertile hatching eggs. Take care to do your homework and find eggs which have been fertilized by a young, healthy, active male duck. Although eggs laid in the dead of winter have been known to hatch successfully, make sure to keep the eggs at a neutral temperature until putting them in the incubator. Most resources recommend storing the eggs at 60-65 degrees. Make sure to keep them with the large end at the top. Although you will need to make arrangements for the eggs well in advance, you may want to wait to actually obtain the eggs until the incubator is up and running. The sooner they can get into the incubator, the better.

Preparing the incubator. Do a dry run of your incubator for at least 24 hours before putting in the hatching eggs. It is important to ensure that your incubator is working correctly and that you are able to maintain the temperature and humidity at a constant rate.

Measuring the temperature at the top of the egg. Although automatic thermometers can be successful in hatching eggs, other times it's important to ensure proper temperature yourself.

The correct temperature for hatching duck eggs is 37.5 degrees Celsius, or 99.5 degrees Fahrenheit.

Mark your fertile hatching eggs with a pencil. If you are turning your eggs by hand, mark them with an X on one side and an O on the other side.

Humidity in the incubator. For duck eggs, keep the humidity at fifty-five percent. This is a little higher than the humidity needed for chicken eggs (40-50 percent).

Turning. Like chicken eggs, duck eggs should be rotated 180 degrees three times a day. Write an X on one side and an O on the other side with a pencil. Three times a day, turn the eggs over from the X side to the O side. Make sure to keep a record of which side you turned them to every morning, afternoon, and evening. If some of the eggs roll around in the incubator and you lose track of when you last turned them, this will help you get all the eggs on the same page.

The Brooder

Once your ducklings hatch, ensure that they have a safe brooder box which is warm, but not too hot.

Temperature in the brooder. Ducklings should be kept at "90-92 degrees for the first 3 days, then 85-90 degrees for days 4 to 7. Thereafter, drop the temperature by approximately 5 degrees per week until they are fully feathered."[320]

Keeping the brooder clean. If you're used to brooding chicks, you're in for a rude awakening. From their first moments of life, ducklings enjoy playing and splashing in water. Unless you're prepared to clean the brooder multiple times a day, you need to develop a system for eliminating water mess and smell.

Puppy pads are a lifesaver for brooding ducklings. Place the puppy pad on the bottom of the brooder to absorb all the poop and water they will splash all over the place!

In addition, it's important to provide plenty of water for the ducklings. They will swim, splash, and play in it, so try to keep your water bowl mostly covered. At the same time, you'll want to provide plenty of water, as they drink a lot, even as new babies.

A good solution to the water problem is a cheap Tupperware container with the corner of the lid cut off. Fill the Tupperware container with water and put the lid on. This allows the ducklings to dunk their heads in the water but does not allow them to swim in it!

Feeding your new duckling. Manna Pro Duck Starter Grower Crumble—this non-medicated feed for young ducks supports healthy digestion."[321]

[320] "How to Raise Your Baby Ducks and Geese," *Grange Coop,* Last modified on August 29, 2015, Accessed on May 21, 2021, https://www.grangecoop.com/how-to-raise-your-baby-ducks-and-geese/.

[321] "Manna Pro Duck Starter Grower Crumble | Non-Medicated Feed for Young Ducks | Supports Healthy Digestion," Amazon, Accessed May 20, 2021, https://www.amazon.com/Manna-Pro-Starter-Grower-Crumble/dp/B0793FBDVJ/ref=as_li_ss_tl?dchild=1&keywords=duck+crumble+feed&qid=1600980258&s=pet-supplies&sr=1-5&linkCode=ll1&tag=thehappychi0c-20&linkId=ec74e2f8f072e3d6b51a8fa400830ad3&language=en_US.

Chapter 9: FAQS

What are some of the most common duck diseases I should be aware of?

Although ducks are generally healthy and easy to care for, sometimes they can fall prey to a variety of duck diseases. Some of the most common diseases are bumblefoot and wet feather.

Bumblefoot. Bumblefoot first shows up as a black spot, a wart-like growth, or another unusual bump on the bird's foot. Bumblefoot originates from cuts and scratches on ducks' tender webbed feet. To avoid bumblefoot, make sure to keep your duck's cage free of pointy objects, rough bedding, and other obstructions. To treat bumblefoot, check out the following article: https://vetericyn.com/blog/signs-and-symptoms-of-bumblefoot-in-chickens.

Wet feather. Wet feather is a common condition that you need to know about. When ducks have a blocked oil gland, nutrient deficit, or other condition that affects the feathers, the self-lubricating mechanism is interrupted, the feathers fray and deteriorate, and the feathers are no longer waterproof. To diagnose your duck with wet feather, check whether water runs off the duck "like water off a duck's back." If your duck looks sopping wet or dirty, your duck may have wet feather. To treat wet feather, give your duck a bath in dish soap! To learn more about treating wet feather, check out this article: https://vetericyn.com/blog/signs-and-symptoms-of-bumblefoot-in-chickens.

How to treat a cut or wound. What if your duck is attacked by a coyote, raccoon, or worse, your own dog? Don't lose hope. Many times, ducks' wounds can be effectively treated. We've seen some pretty terrible wounds that have healed up just fine. Apply antibacterial ointment and keep the animal in a quiet, safe place in isolation. Use sterile wrap to keep the affected area clean.

How much time per day do ducks require for their care?

If you're accustomed to raising chickens, you'll need to get used to spending more time outside with your ducks. In my experience, the difference was astronomical and exponential.

At the same time, as I've gotten more accustomed to raising ducks, the amount of time I spend on them has dropped. I feed them in the morning in less than fifteen minutes. I refill their water pool every day or every few days. With better drainage and higher quality bedding, I've reduced the amount of time I spend cleaning their pen as well. Sometimes, I can go several weeks without changing the bedding!

Why is my duck losing all her feathers?

Once a year, ducks molt. This means that they lose all their feathers and grow new ones. If your duck has stopped laying and is losing her feathers, don't panic! This is normal. When the duck finishes molting, she will have an all-new winter coat! In addition, molting often takes care of wet feather, so if your duck has wet feather, you may need to do nothing; just wait for the next molt!

Why did my ducks refuse to swim in their pool?

Ducks love water, but they are also highly finicky creatures. Your duck may refuse to swim because he or she had a bad experience in the past. For example, as mentioned above, your duck may have had a difficult time getting into or out of the pool in the past. There may not have been a gently sloping ramp that helped your duck get in. The "plop" into the water or the "plop" back out onto dry land may have scared your duck, and he may be hesitant to try again.

Further, your duck may be experiencing a disease, such as wet feather, that makes him or her reluctant to swim. A duck that has wet feather does not stay waterproof while swimming. In winter or in cold, drafty weather, this duck will feel soaked to the bone. It will become easily chilled. Knowing how cold it feels when it swims, it may avoid swimming in the future.

To encourage your duck to swim, try adding one of its favorite foods to the water. Do you have a kiddie pool that has a big enough diameter that your duck can't reach the middle while standing at the side? Try tossing in some snacks or some wheat to the middle of your pool. Your duck won't be able to reach it from the side, but his greedy nature will encourage him to jump right in so he can reach it. Any time you can add some wheat, grain, or other yummy snacks to the water, your duck will become more likely to submerge his head, splash, and play in the water.

Where can I meet other duck owners, get support, and ask questions?

Social media duck groups are excellent places to ask questions, let off steam, and get some comic relief. When I first joined a popular duck Facebook group, I found I spent so much time just scrolling and chuckling. I could identify so strongly with these comical posts. Ducks are such funny creatures, and duck owners love to share stories with others who have the same interests.

How can I keep my ducks from eating their own eggs?

Use ceramic eggs or golf balls to discourage your ducks from eating their eggs. Gather eggs frequently. Don't give them the chance to snack on eggs. Bake or cook eggshells before returning them to the ducks as a calcium supplement. In this way, the ducks will avoid developing a taste for their own raw eggs.

Why is the albumen (egg white) bright green?

According to *This NZ Life,* a green albumen "can mean an excess of riboflavin in the diet, or that a bird has eaten a weed called shepherd's purse."[322]

Why is there blood in my duck's egg?

Blood in the egg could indicate that your duck has been brooding and her eggs have started to develop. Don't eat a bloody egg.

Why does my duck egg have two yolks?

Sometimes, ducks lay double-yolk eggs. This phenomenon is not common, but it is not abnormal. It is similar to a human having twins. These eggs do not hatch often, but in very rare occasions with very intensive care, they can hatch.

---

[322] Sue Clarke, "5 common egg imperfections and what they say about your chickens," *This NZ Life,* Accessed on May 21, 2021, https://thisnzlife.co.nz/5-common-egg-imperfections-and-what-they-say-about-your-chickens/.

Bibliography

"About Us and Our Faith," JM Hatchery, Accessed April 27, 2021, https://jmhatchery.com/about-us-and-our-faith/.

"Ammonia Causes and Effects," *Metzer Farms,* Last modified on December 1, 2017, Accessed on May 19, 2021, https://metzerfarms.blogspot.com/2017/12/ammonia-causes-and-affects.html.

"Bantam Ducks Make Good Pets," *Ashton Waterfowl,* Accessed April 26, 2021, https://ashtonwaterfowl.net/bantam_ducks.htm.

"Best Meat Duck Breeds," *The Happy Chicken Coop,* Last modified on November 24, 2020, Accessed April 26, 2021, https://www.thehappychickencoop.com/best-meat-duck-breeds/.

"Black East Indian Ducks," *Poultry Keeper,* Accessed April 26, 2021, https://poultrykeeper.com/duck-breeds/black-east-indian-ducks/.

"Blue Swedish Duck," *McMurray Hatchery,* Accessed April 26, 2021, https://www.mcmurrayhatchery.com/blue_swedish.html.

"Blue Swedish Ducks (Anas platyrhynchos domesticus)," *Beauty of Birds,* Accessed April 26, 2021, https://www.beautyofbirds.com/blueswedishducks.html.

"Call Duck," *Omlet,* Accessed March 30, 2021, https://www.omlet.us/breeds/ducks/call_duck/.

"Call Ducks," *Wikipedia,* Accessed April 26, 2021, https://en.wikipedia.org/wiki/Call_duck.

"Carrie," *Farm Sanctuary,* Accessed April 27, 2021, https://www.farmsanctuary.org/adopt/adopt-a-farm-animal-carrie/.

"Cheap DIY Bio Pond Filter that WORKS!!!," Last modified on November 20, 2011, Accessed on May 19, 2021, YouTube video, 3:20, https://www.youtube.com/watch?v=VMmvKpNrcTg&ab_channel=DavidEkstrom.

"Commercial Poultry Production Air Temperature," *University of Kentucky College of Agriculture, Food and Environment,* Accessed March 29, 2021, https://afs.ca.uky.edu/poultry/chapter-7-air-temperature#:~:text=In%20the%20adult%20chicken%20the,body%20temperature%20than%20larger%20breeds.

"Do It Yourself Bio Filter," Accessed May 19, 2021, http://www.runnerduck.com/pf1.htm.

"Duck coop on concrete slab??" *Backyard Chickens,* Last modified on May 31, 2011, Accessed on May 7, 2021, https://www.backyardchickens.com/threads/duck-coop-on-concrete-slab.515211/.

"Gamma2 Vittles Vault Airtight Pet Food Storage Container," Amazon Description, Accessed April 26, 2021, https://www.amazon.com/Vittles-Vault-Outback-Airtight-Container/dp/B0002DJOOI/ref=sr_1_3?dchild=1&gclid=Cj0KCQjwyZmEBhCpARIsALIzmnLhalmiUz0dX0W4Ypt8xu3Os8UkX9ieuNzFK_4ajx8fQZeU6vFu7ccaAlS7EALw_wcB&hvadid=234476390778&hvdev=c&hvlocphy=1017673&hvnetw=g&hvqmt=e&hvrand=16883132225262277533&hvtargid=kwd-298629170360&hydadcr=24961_10366456&keywords=vittles+vault&qid=1619463221&s=pet-supplies&sr=1-3.

"Golden Cascade Duck," Last modified on July 3, 2016, Accessed on April 27, 2021, https://www.breedslist.com/golden-cascade-duck.htm.

"Golden Cascade Duck: Characteristics, Uses & Full Breed Information," *Roy's Farm,* Last modified on March 11, 2021, Accessed April 27, 2021, https://www.roysfarm.com/golden-cascade-duck/.

"Golden Cascade," *Wikipedia,* Accessed April 27, 2021, https://en.wikipedia.org/wiki/Golden_Cascade.

"Great Backyard Duck Breeds," Accessed April 26, 2021, *The Cape Coop,* https://thecapecoop.com/great-backyard-duck-breeds/.

"Growing Fodder for Backyard Chickens, Ducks, and Geese," *Fresh Eggs Daily,* Accessed May 21, 2021, https://www.fresheggsdaily.blog/2014/02/growing-sprouted-fodder-for-your.html.

"How to Conduct a Duck Health Examination," Last modified on April 10, 2020, Accessed on May 6, 2021, https://opensanctuary.org/article/how-to-conduct-a-duck-health-examination/.

"How to Make a Duck House," *The Cape Coop,* Accessed May 19, 2021, https://thecapecoop.com/make-duck-house/.

"How to Save Money by Fermenting Chicken Feed," *Fresh Eggs Daily,* Accessed May 21, 2021, https://www.fresheggsdaily.blog/2015/03/how-to-save-money-fermenting.html.

"INTRODUCING THE "QUACKER BOX" DUCK HOUSE – A BIRTHDAY PRESENT WORTHY OF THE TYRANT," Accessed May 19, 2021, Last modified on September 27, 2018, https://www.tyrantfarms.com/introducing-the-quaker-box-a-birthday-present-worthy-of-the-tyrant/.

"Mallards," *National Geographic,* Accessed April 26, 2021, https://www.nationalgeographic.com/animals/birds/facts/mallard.

"Manna Pro All Flock Crumbles | 16% Protein Level | Complete Feed for Chickens, Ducks, Geese, Turkeys and Gamebirds | Probiotics to Support Digestion | Crumbled Form for Easy Feeding | 25 Pounds," Amazon, Accessed May 20, 2021, https://www.amazon.com/Manna-Pro-Probiotics-Formulated-Vitamins/dp/B01N4KD9MY/ref=as_li_ss_tl?ie=UTF8&aaxitk=IsJCQ8u0tc8rQ5KfMAdJxA&hsa_cr_id=5173855720401&pd_rd_r=f115d2e9-e817-4147-bdbd-44962adc6921&pd_rd_w=i9eeZ&pd_rd_wg=jXHsY&ref_=sbx_be_s_sparkle_mcd_asin_1_img&linkCode=ll1&tag=thehappychi0c-20&linkId=785c599368cac511a4c86f7fc69e1375&language=en_US.

"Manna Pro Duck Starter Grower Crumble | Non-Medicated Feed for Young Ducks | Supports Healthy Digestion," Amazon, Accessed May 20, 2021, https://www.amazon.com/Manna-Pro-Starter-Grower-Crumble/dp/B0793FBDVJ/ref=as_li_ss_tl?dchild=1&keywords=duck+crumble+feed&qid=1600980258&s=pet-supplies&sr=1-5&linkCode=ll1&tag=thehappychi0c-20&linkId=ec74e2f8f072e3d6b51a8fa400830ad3&language=en_US.

"Nutritional Requirements For Ducks and Geese," *Metzer Farms,* Accessed May 21, 2021, https://www.metzerfarms.com/NutritionalRequirements.cfm.

"Pekin Duck Breed: Everything You Need to Know," *Happy Chicken,* Last modified November 3, 2020, Accessed April 26, 2021, https://www.thehappychickencoop.com/pekin-duck-breed-everything-you-need-to-know/.

"Preschooler Screen Time Linked to Attention Problems," *Cleveland Clinic,* Last modified on July 18, 2019, Accessed on March 30, 2021, https://newsroom.clevelandclinic.org/2019/07/18/preschooler-screen-time-linked-to-attention-problems/.

"Saxony Duck," Burke's Backyard, Accessed March 30, 2021, https://www.burkesbackyard.com.au/fact-sheets/pets/pet-road-tests/saxony-duck/.

"Self-Draining Duck Pen—Is It Possible?" *Backyard Chickens,* Lost modified on April 11, 2014, Accessed on May 7, 2021, https://www.backyardchickens.com/threads/self-draining-duck-pen-is-it-possible.872928/.

"Silky Duck," *Holderread Farm,* Accessed April 27, 2021, https://www.holderreadfarm.com/photogallery/silkies_page/silkies_page.htm.

"Silky Ducks," *Feathersite,* Accessed April 27, 2021, https://www.feathersite.com/Poultry/Ducks/Silky/BRKSilky.html.

"Standard Colours of the Indian Runner Duck," *Indian Runner Duck Club,* Accessed April 26, 2021, https://www.runnerduck.net/standard-colours.php.

"Storey's Guide to Raising Ducks, 2nd Edition: Breeds, Care, Health," Amazon Description, https://www.amazon.com/dp/1603426922?tag=happy_aff1-20&linkCode=ogi&th=1&psc=1.

"Swedish Blue," *Wikipedia,* Accessed April 26, 2021, https://en.wikipedia.org/wiki/Swedish_Blue.

"The Silver Appleyard: A Great All-Round Duck," *The Modern Homestead,* Accessed April 26, 2021, https://www.themodernhomestead.us/article/Silver+Appleyard.html.

"Top Duck-Craving Predators," *Delta Waterfowl,* Accessed May 7, 2021, https://deltawaterfowl.org/top-duck-craving-predators/.

"Welsh Harlequin Duck: Everything You Need to Know," *The Happy Chicken Coop,* Last modified on November 12, 2020, Accessed on April 27, 2021, https://www.thehappychickencoop.com/welsh-harlequin-duck-everything-you-need-to-know/.

"What Does Your Duck House Look Like," *Backyard Chickens,* Last modified on March 3, 2010, Accessed on May 19, 2021, https://www.backyardchickens.com/threads/what-does-your-duck-house-look-like.178796/page-2.

"What Does Your Duck House Look Like," *Backyard Chickens,* Last modified on March 3, 2010, Accessed on May 19, 2021, https://www.backyardchickens.com/threads/what-does-your-duck-house-look-like.178796/page-3.

"What not to feed chickens and ducks!" *Under a Tin Roof,* Accessed May 20, 2021, https://www.underatinroof.com/blog/2018/6/14/what-not-to-feed-chickens-and-ducks.

"What to expect at a chicken swap (you might be surprised!)" *Murano Chicken Farm,* Accessed April 27, 2021, https://www.muranochickenfarm.com/2014/05/what-to-expect-at-chicken-swap.html.

"Winter Duck Care," *The Cape Coop,* Accessed March 29, 2021, https://thecapecoop.com/winter-duck-care/.

Aaron and Susan. "DUCK EGGS VS. CHICKEN EGGS: HOW DO THEY COMPARE?" *Tyrant Farms,* Last updated on January 13, 2019. https://www.tyrantfarms.com/5-things-you-didnt-know-about-duck-eggs/.

Aaron and Susan. "How To Get Your Ducks to Like You: Three Tips." *Tyrant Farms,* Last modified on October 6, 2017. https://www.tyrantfarms.com/how-to-get-your-ducks-to-like-you-three-tips/.

"Abacot Ranger Duck," *Raising Ducks,* Accessed April 26, 2021, https://www.raising-ducks.com/duck-breed-guide/abacot-ranger-duck/.

"Abacot Ranger Ducks," *Poultry Keeper,* Accessed April 26, 2021, https://poultrykeeper.com/duck-breeds/abacot-ranger-ducks/.

April Lee, "9 Quietest Duck Breeds," Accessed April 26, 2021, https://farmhouseguide.com/quietest-duck-breeds/.

"Aylesbury Duck," *Heritage Poultry,* Accessed April 26, 2021, https://heritagepoultry.org/blog/aylesbury-duck.

"Aylesbury Duck," *Wikipedia,* Accessed April 26, 2021, https://en.wikipedia.org/wiki/Aylesbur.

"Bali Ducks," *Poultry Keeper,* Accessed April 26, 2021, https://poultrykeeper.com/duck-breeds/bali-ducks/.

"Bali," *Omlet,* Accessed April 26, 2021, https://www.omlet.us/breeds/ducks/bali/.

Bill Chappell, "'We Are Swamped': Coronavirus Propels Interest In Raising Backyard Chickens For Eggs," Last modified on April 3, 2020, Accessed on May 6, 2021, https://www.npr.org/2020/04/03/826925180/we-are-swamped-coronavirus-propels-interest-in-raising-backyard-chickens-for-egg.

"Black Swedish Duck," *Murray McMurray,* Accessed April 27, 2021, https://www.mcmurrayhatchery.com/black_swedish_duck.html.

"Ducklings: Black Swedish," *My Pet Chicken,* Accessed April 27, 2021, https://www.mypetchicken.com/catalog/Waterfowl/Ducklings-Black-Swedish-p2565.aspx.

"Black Swedish Ducklings," *Purely Poultry,* Accessed April 27, 2021, https://www.purelypoultry.com/black-swedish-ducklings-p-862.html.

"Breed Profile: Cayuga Duck," *Backyard Poultry,* Accessed April 26, 2021, https://backyardpoultry.iamcountryside.com/poultry-101/cayuga-duck-breed-spotlight/.

"Buff Orpington Duck Breed: Everything You Need To Know," *The Happy Chicken Coop,* Accessed April 26, 2021, https://www.thehappychickencoop.com/buff-orpington-duck-breed-everything-you-need-to-know/.

"Cayuga Duck," *The Livestock Conservancy,* Accessed April 26, 2021, https://livestockconservancy.org/index.php/heritage/internal/cayuga.

Clarke, Sue, "5 common egg imperfections and what they say about your chickens," *This NZ Life,* Accessed on May 21, 2021, https://thisnzlife.co.nz/5-common-egg-imperfections-and-what-they-say-about-your-chickens/.

"Crested (duck breed)," *Wikipedia,* Accessed April 26, 2021, https://en.wikipedia.org/wiki/Crested_(duck_breed).

"Crested Duck: Characteristics, Origin, Uses & Full Breed Information," *Roy's Farm,* Last modified March 11, 2021, Accessed on April 26, 2021, https://www.roysfarm.com/crested-duck/.

"Crested Miniature," *Omlet,* Accessed April 26, 2021, https://www.omlet.us/breeds/ducks/crested_miniature/.

"Crested Miniature Ducks," *Poultry Keeper,* Accessed April 26, 2021, https://poultrykeeper.com/duck-breeds/crested-miniature-ducks/.

D Holderread, H S Nakaue, G H Arscott, "Anticoccidial drugs and duckling performance to four weeks of age," 1983 Jun; 62(6) : 1125-7. doi: 10.3382 /ps.0621125, Accessed May 20, 2021, https://pubmed.ncbi.nlm.nih.gov/6878147/.

Erica, "Honest reasons you shouldn't get ducks," *Northwest Edible Life,* Last modified on May 17, 2017, Accessed March 29, 2021, https://nwedible.com/reasons-to-not-get-ducks/.

"F1 Hybrid," *Wikipedia,* Accessed April 27, 2021, https://en.wikipedia.org/wiki/F1_hybrid.

"Foie Gras," *Wikipedia,* Accessed April 27, 2021, https://en.wikipedia.org/wiki/Foie_gras.

Hana LaRock, "How to Keep a Duck Pond Clean Naturally," *SF Gate,* Accessed May 19, 2021, Last modified on October 8, 2019, https://homeguides.sfgate.com/keep-duck-pond-clean-naturally-84449.html.

Henke, Jodi, "Raising Welsh Harlequin Ducks," *Successful Farming,* Last modified on November 25, 2013, Accessed on April 27, 2021, https://www.agriculture.com/family/living-the-country-life/raising-welsh-harlequin-ducks.

"Hook Bill Ducks," *Poultry Keeper,* Accessed April 26, 2021, https://poultrykeeper.com/duck-breeds/hook-bill-ducks/.

"How to Raise Your Baby Ducks and Geese," *Grange Coop,* Last modified on August 29, 2015, Accessed on May 21, 2021, https://www.grangecoop.com/how-to-raise-your-baby-ducks-and-geese/.

https://www.underatinroof.com/blog/2018/6/14/what-not-to-feed-chickens-and-ducks

https://www.understood.org/en/learning-thinking-differences/child-learning-disabilities/sensory-processing-issues/heavy-work-activities

"Indian Runner Duck," *Oregon Zoo,* Accessed April 26, 2021, https://www.oregonzoo.org/discover/animals/indian-runner-duck.

"Indian Runner Duck," *Wikipedia,* Accessed April 26, 2021, https://en.wikipedia.org/wiki/Indian_Runner_duck.

"Information about Ducks to Keep them Warm and Healthy Through the Winter," *Backyard Poultry,* Last modified on October 26, 2020, Accessed May 19, 2021, https://backyardpoultry.iamcountryside.com/coops/duck-shelters-for-winter/.

Jason Roberts, "17 Best Hatcheries to Buy Chickens Online," Last modified on April 25, 2020, Accessed on May 6, 2021, https://www.knowyourchickens.com/buy-chickens-online.

Jennifer Poindexter, "37 Free DIY Duck House / Coop Plans & Ideas that You Can Easily Build," *Morning Chores,* Accessed May 19, 2021, https://morningchores.com/duck-house-plans/.

K J Theodore, "Common Poultry Myths," *Shagbark Bantams,* Accessed May 20, 2021, https://www.shagbarkbantams.com/common-poultry-myths/.

Kennedy, Pagan. "How to Get High on Soil." *The Atlantic,* Last modified on January 31, 2012. https://www.theatlantic.com/health/archive/2012/01/how-to-get-high-on-soil/251935/.

"Khaki Campbell Duck," *Livestock Conservancy,* Accessed April 26, 2021, https://livestockconservancy.org/index.php/heritage/internal/campbell.

"Khaki Campbell Duck: Everything You Need To Know," *The Happy Chicken Coop,* Last modified on October 27, 2020, Accessed April 26, 2021, https://www.thehappychickencoop.com/khaki-campbell-duck-everything-you-need-to-know/.

Kim Irvine, "Appleyard Duck Breed – Everything You Need to Know," Last modified on November 26, 20218, *Domestic Animal Breeds,* Accessed April 26, 2021, https://domesticanimalbreeds.com/appleyard-duck-breed-everything-you-need-to-know/.

Kim Johnson, "What is Foie Gras," *Animal Equality,* Last modified on July 26, 2019, Accessed on April 27, 2021, https://animalequality.org/blog/2019/07/26/what-is-foie-gras/.

"KOOP CLEAN™ CHICKEN BEDDING: Superior Chicken Bedding for a Fresh, Happy Flock," *Lucerne Farms,* Accessed May 21, 2021, https://lucernefarms.com/koop-clean/.

Lisa Steele, "A Guide to Duck Houses," *HGTV,* Accessed on May 7, 2021, https://www.hgtv.com/outdoors/gardens/animals-and-wildlife/a-guide-to-duck-houses.

Lisa Steele, "Can Chickens & Ducks Share Feed?," *Hobby Farms,* Last modified on January 2, 2017, Accessed on May 20, 2021, https://www.hobbyfarms.com/can-chickens-ducks-share-feed/.

Lisa Steele, "A Guide to Duck Shelters for Winter: *Information about Ducks to Keep them Warm and Healthy Through the Winter," Backyard Poultry,* Last modified on October 26, 2020, Accessed May 21, 2021, https://backyardpoultry.iamcountryside.com/coops/duck-shelters-for-winter/.

"Magpie Duck," *The Livestock Conservancy*, Accessed April 26, 2021, https://livestockconservancy.org/index.php/heritage/internal/magpie.

"Magpie Duck," *Wikipedia,* accessed April 26, 2021, https://en.wikipedia.org/wiki/Magpie_duck.

"Magpie Duck: Characteristics, Origin, Uses & Full Breed Information," *Roys Farms,* Accessed April 26, 2021, https://www.roysfarm.com/magpie-duck/.

"Mallard," *Wikipedia,* Accessed April 26, 2021, https://en.wikipedia.org/wiki/Mallard.

"Manna Pro Duck Layer Pellet | High Protein for Increased Egg Production | Formulated with Probiotics to Support Healthy Digestion | 8 Pounds," Amazon description, Accessed May 20, 2021, https://www.amazon.com/Manna-Pro-Duck-Layer-Pellets/dp/B0793FPLMR/ref=as_li_ss_tl?dchild=1&keywords=duck+crumble+feed&qid=1600980258&s=pet-supplies&sr=1-2-spons&psc=1&spLa=ZW5jcnlwdGVkUXVhbGlmaWVyPUEzM1BJOFJVNjVXVlQoJmVuY3J5cHRlZElkPUEwNDg5MzgzVVc2M0xHS1c2TlVZJmVuY3J5cHRlZEFkSWQ9QTAwNTUwMDYyQTFVRjBCCNU9NU1I4JndpZGdldE5hbWU9c3BfYXRmJmFjdGlvbj1jbGlja1JlZGlyZWN0Jm5vTG9nQ2xpY2s9dHJ1ZQ==&linkCode=ll1&tag=thehappychi0c-20&linkId=66e6667e9aab6b18d56918be94b7c477&language=en_US.

Mary, "How to Get Rid of Flies in Your Chicken Coop, Naturally," *Life is Just Ducky,* Accessed April 26m 2021, https://www.lifeisjustducky.com/how-to-get-rid-of-flies/.

Melissa Mayntz, "What do Ducks and Other Wildfowl Eat?," Last modified on December 23, 2020, Accessed May 20, 2021, https://www.thespruce.com/what-to-feed-ducks-386584.

Morin, Amanda. "Heavy Work and Sensory Processing Issues: What You Need to Know." *Understood,* Accessed March 29, 2021, https://www.understood.org/en/learning-thinking-differences/child-learning-disabilities/sensory-processing-issues/heavy-work-activities.

"Mulard," *Wikipedia,* Accessed April 27, 2021, https://en.wikipedia.org/wiki/Mulard.

"Muscovy Duck," *Wikipedia,* Accessed April 27, 2021, https://en.wikipedia.org/wiki/Muscovy_duck.

"Muscovy Duck: Eggs, Facts, Care Guide and More..." *The Happy Chicken Coop,* Accessed April 27, 2021, https://www.thehappychickencoop.com/muscovy-duck/.

Nosowitz, Dan. "Everything You Need to Know About Duck Eggs." *Modern Farmer.* Last modified on June 19, 2015.

https://modernfarmer.com/2015/06/everything-you-need-to-know-about-duck-eggs/.

"Orpington Duck," *Wikipedia,* Accessed April 26, 2021, https://en.wikipedia.org/wiki/Orpington_Duck.

"Parasites—Liver Flukes," *Centers for Disease Control and Prevention,* Accessed April 26, 2021, https://www.cdc.gov/parasites/liver_flukes/index.html.

"Pekin Duck," *Columbian Park Zoo,* Accessed April 27, 2021, http://www.lafayette.in.gov/DocumentCenter/View/1806/Pekin-Duck-PDF#:~:text=Physical%20Characteristics%3A%20Weigh%207%2D9,Ducklings%20are%20yellow.

"Poultry Breeds—Rouen Duck," *Breeds of Livestock, Department of Animal Science,* Accessed April 27, 2021, http://afs.okstate.edu/breeds/poultry/ducks/rouen/index.html/.

"Rouen Clair," *Omlet,* Accessed April 27, 2021, https://www.omlet.us/breeds/ducks/rouen_clair/.

"Rouen," *Omlet,* Accessed April 27, 2021, https://www.omlet.us/breeds/ducks/rouen/.

"Rouen—Non-industrial Duck," Livestock Conservancy, Accessed April 27, 2021, https://livestockconservancy.org/index.php/heritage/internal/rouen.

Samantha Johnson, "6 Duck Breeds to Raise for Eggs," *Hobby Farms,* Last modified March 15, 2019, Accessed on April 26, 2021, https://www.hobbyfarms.com/6-duck-breeds-to-raise-for-eggs-4/.

Sara Barnes, "Stylish Ducks Waddle Down the Catwalk in Annual Fashion Show," Last modified August 20, 2014, *My Modern Met,* Accessed April 26, 2021, https://mymodernmet.com/australian-pied-piper-duck-show/.

Sarah Moore, "Which Fowl Eat Mosquitoes?" *Pets on Mom.com,* Accessed April 26, 2021, https://animals.mom.com/fowl-eat-mosquitoes-11665.html.

"Silver Appleyard Miniature," *Omlet UK,* Accessed April 26, 2021, https://www.omlet.co.uk/breeds/ducks/silver_appleyard_miniature.

"Silver Appleyard Reviews," *Omlet,* Accessed April 26, 2021, https://www.omlet.us/breeds/ducks/appleyard/reviews.

"Silver Bantam Duck," *Raising Ducks,* Accessed April 26, 2021, https://afowlshome.com/types-of-fowl/ducks/domestic-ducks/bantam-ducks/silver-bantam-ducks/.

Snyder, Cecilia, MS, RD. "Duck Eggs vs. Chicken Eggs: Nutrition, Benefits, and More." Last modified on December 1, 2020. https://www.healthline.com/nutrition/duck-eggs-vs-chicken-eggs#nutritional-comparison.

"The Best Livestock Feed Storage Containers," *Simple Living Country Gal,* Accessed April 26, 2021, https://simplelivingcountrygal.com/livestock-feed-storage-containers/.

"The Cayuga Duck Originates From New York, Named After Lake Cayuga," *Wide Open Pets,* Accessed April 26, 2021, https://www.wideopenpets.com/cayuga-duck/.

Tim Daniels, "Feeding Ducks," *Poultry Keeper,* Last updated on September 11, 2018, Accessed May 20, 2021, https://poultrykeeper.com/duck-keeping/feeding-ducks/.

Walter Jeffries, "Gold Star Ducks," *Sugar Mountain Farm,* Last modified on August 12, 2013, Accessed on April 27, 2021, https://sugarmtnfarm.com/2013/08/12/gold-star-ducks/.

"Welsh Harlequin Duck," *Murray McMurray,* Accessed April 27, 2021, https://www.mcmurrayhatchery.com/welsh_harelquin_duck.html.

"What color is it?" *The Ugly Duck Farm,* Accessed April 27, 2021, http://muscovy.us/what_color_is_it.

"White Star Hybrid," Murray McMurray Hatchery, Accessed April 27, 2021, https://www.mcmurrayhatchery.com/white_star_hybrid_white_layer_ducks.html.

Yang, Jianmei, Hongrui Cui, Qiaoyang Teng, Wenjun Ma, Xuesong Li, Binbin Wang, Dawei Yan, Hongjun Chen, Qinfang Liu and Zejun Li. "Ducks induce rapid and robust antibody responses than chickens at early time after intravenous infection with H9N2 avian influenza virus." *Virology Journal.* Last modified on

April 11, 2019. https://virologyj.biomedcentral.com/articles/10.1186/s12985-019-1150-8.